Springer Theses

Recognizing Outstanding Ph.D. Research

Aims and Scope

The series "Springer Theses" brings together a selection of the very best Ph.D. theses from around the world and across the physical sciences. Nominated and endorsed by two recognized specialists, each published volume has been selected for its scientific excellence and the high impact of its contents for the pertinent field of research. For greater accessibility to non-specialists, the published versions include an extended introduction, as well as a foreword by the student's supervisor explaining the special relevance of the work for the field. As a whole, the series will provide a valuable resource both for newcomers to the research fields described, and for other scientists seeking detailed background information on special questions. Finally, it provides an accredited documentation of the valuable contributions made by today's younger generation of scientists.

Theses may be nominated for publication in this series by heads of department at internationally leading universities or institutes and should fulfill all of the following criteria

- They must be written in good English.
- The topic should fall within the confines of Chemistry, Physics, Earth Sciences, Engineering and related interdisciplinary fields such as Materials, Nanoscience, Chemical Engineering, Complex Systems and Biophysics.
- The work reported in the thesis must represent a significant scientific advance.
- If the thesis includes previously published material, permission to reproduce this must be gained from the respective copyright holder (a maximum 30% of the thesis should be a verbatim reproduction from the author's previous publications).
- They must have been examined and passed during the 12 months prior to nomination.
- Each thesis should include a foreword by the supervisor outlining the significance of its content.
- The theses should have a clearly defined structure including an introduction accessible to new PhD students and scientists not expert in the relevant field.

Indexed by zbMATH.

Kristian Blom

Pair-Correlation Effects in Many-Body Systems

Towards a Complete Theoretical Description of Pair-Correlations in the Static and Kinetic Description of Many-Body Systems

Doctoral Thesis accepted by
Georg-August-Universität Göttingen, Göttingen, Germany

Springer

Author
Dr. Kristian Blom
Department of Theoretical
and Computational Biophysics
Max Planck Institute for Multidisciplinary
Sciences
Göttingen, Niedersachsen, Germany

Supervisor
Dr. Aljaž Godec
Head of the Mathematical Biophysics
Group
Max Planck Institute for Multidisciplinary
Sciences
Göttingen, Niedersachsen, Germany

ISSN 2190-5053 ISSN 2190-5061 (electronic)
Springer Theses
ISBN 978-3-031-29611-6 ISBN 978-3-031-29612-3 (eBook)
https://doi.org/10.1007/978-3-031-29612-3

This Springer imprint is published by the registered company Springer Nature Switzerland AG
The registered company address is: Gewerbestrasse 11, 6330 Cham, Switzerland

To Mareile, who is always there for me.

Supervisor's Foreword

Notwithstanding the efforts and remarkable advances put forward during the last century, our understanding of strongly correlated many-body systems is still very limited, especially when considering inhomogeneous systems or systems driven out of thermodynamic equilibrium. This is because exact results in dimensions larger than one are very difficult (if not impossible) to obtain, and even in cases when they can be derived, it is often very difficult to interpret them in a physically intuitive manner. The reason lies in the treatment of correlations between individual particles, which may have highly non-trivial and, as Kristian Blom discovered in his thesis, quite unexpected but critical implications for the thermodynamics and kinetics of many-body systems. It is therefore natural on the one hand, and required on the other hand, to resort to approximations. Often the underlying complexity seems to imply that one must stop already at the lowest, mean-field level, and even this frequently turns out to be a difficult task, especially in a non-equilibrium setting.

A lot of effort has therefore been devoted to constructing better approximate methods—those that take into account correlations between particles and allow for accurate analytical solutions. One such accurate method is the so-called Bethe–Guggenheim theory that was developed independently by Hans Bethe and Edward Guggenheim in 1935 and explicitly takes into account nearest-neighbor pair correlations. However, albeit powerful and accurate, the combinatorics within the Bethe–Guggenheim theory is already substantially more challenging as compared to mean-field theory. Therefore, to date Bethe–Guggenheim theory was essentially limited to the treatment of spatially homogeneous systems at thermodynamic equilibrium.

In view of the rapid recent developments in non-equilibrium statistical physics, which include far-from-equilibrium fluctuation-dissipation relations, anomalous relaxation phenomena (a.k.a. the "Mpemba effect"), frenetic concepts, asymmetries and dynamical phase transitions in non-equilibrium relaxation, as well as thermodynamic uncertainty relations and speed limits, classical many-body systems in dimensions 2 and 3 seem to be a particularly "tough nut to crack".

From this point, being confined solely to the realm of mean-field results, the so-called random-phase approximation, self-consistent field theory, and diverse field-theoretic approaches that are valid near the critical point are still unsatisfactory. This is especially so, because these techniques either do not apply to non-uniform systems, or are applicable in a very limited range of interaction strengths.

In his thesis, Kristian Blom embarked on the conceptually and technically challenging task of generalizing Bethe–Guggenheim theory to describe and investigate the importance of pair correlations for phenomena in spatially inhomogeneous (in general strongly interacting) classical many-body systems at equilibrium and for the kinetics of such systems when these are brought out of thermodynamic equilibrium. His microscopic starting point thereby was throughout the paradigmatic nearest-neighbor interacting Ising model in dimensions above one, evolving according to Glauber or (magnetization-conserving) Kawasaki dynamics.

Kristian Blom creatively combines a variety of advanced methods, including asymptotic analysis, reversion of series, the squeeze theorem, perturbation theory, large deviation theory, and a mapping of spin dynamics onto analytical mechanics, which are presented explicitly and in a highly pedagogical manner. The analytical "tour de force" is throughout corroborated by computer simulations of the Ising model and discussed in the context of relevant experiments.

The thesis significantly advances our understanding of the thermodynamics and kinetics of stochastic many-body systems from both technical and conceptual points of view. Three aspects deserve to be pointed out in particular. First is the unprecedented, fully analytical account of the process of magnetization reversal in the Ising ferromagnet and its application in the biological context of cell (de)adhesion, which uncovered an unforeseen kind of dynamical phase transition, where the statistically dominant transition path undergoes an abrupt qualitative change upon varying the interaction strength. Second is the derivation of a thermodynmically consistent phase-field theory that accounts for nearest-neighbor pair correlation effects, by evaluating explicitly the thermodynamic limit of a spatially inhomogeneous Ising model. And third, the discovery and explanation of a global speed limit to relaxation kinetics in far-from-equilibrium temperature quenches, which are clearly shown to require accounting for pair correlations.

The general importance of thermodynamic consistency of non-equilibrium, in particular far-from-equilibrium, theories is meanwhile fully appreciated, as is accounting for correlations between particles in many-body physics. However, a consistent combination of both is to a large extent still elusive, whereas many important questions remain unanswered, and the results put forward in the thesis of Kristian Blom provide a powerful means to account for particle correlations in an accurate, comprehensible, tractable, and thermodynamically consistent manner that has the potential to shape future developments in the field.

Göttingen, Germany
February 2023

Dr. Aljaž Godec

Abstract

The laws of nature encompass the small, the large, the few, and the many. In this thesis, we are concerned with classical (i.e., not quantum) many-body systems, which refers to any microscopic or macroscopic system that contains a large number of interacting entities. The nearest-neighbor Ising model, originally developed in 1920 by Wilhelm Lenz, forms a cornerstone in our theoretical understanding of collective effects in classical many-body systems and is to date a paradigm for statistical physics.

Despite its elegant and simplistic description, exact analytical results in dimensions equal and larger than two are difficult to obtain. Therefore, much work has been done to construct methods that allow for approximate, yet accurate, analytical solutions. One of these methods is the Bethe–Guggenheim approximation, originally developed independently by Hans Bethe and Edward Guggenheim in 1935. This approximation goes beyond the well-known mean-field approximation and explicitly accounts for pair correlations between the spins in the Ising model. In this thesis, we embark on a journey to exploit the full capacity of the Bethe–Guggenheim approximation, in non-uniform and non-equilibrium settings.

After we formally introduce the original Bethe–Guggenheim approximation for uniform systems, we will extend its scope to non-uniform systems and derive a Cahn-Hilliard free energy functional. Here we find that the one-dimensional equilibrium concentration profile undergoes a delocalization-induced broadening transition at interaction strengths near and above the thermal energy. The broadening transition arises from a decreasing amplitude of capillary wave fluctuations and is not accounted for in the mean-field approximation.

Finally, going beyond equilibrium properties, we also study the kinetics of the Ising model in and out of equilibrium. First, based on firm theoretical and experimental evidence, we construct an Ising-like minimal model for interacting cellular adhesion molecules. Here we find a dynamical critical point where adhesion cluster dissolution and formation are the fastest and undergo a qualitative change in dynamics. Second, we demonstrate the existence of a finite-time dynamical phase transition for disordering quenches in the nearest-neighbor Ising model. Starkly different from the mean-field approximation, the time at which the dynamical phase transition occurs is bounded from below by a speed limit.

Altogether, in this thesis, we unveil the non-trivial and a priori non-intuitive effects of pair correlations in the classical nearest-neighbor Ising model, which are taken into account in the Bethe–Guggenheim approximation, and neglected in the mean-field approximation.

Acknowledgements

While pursuing a Ph.D. in the *Mathematical bioPhysics Group* in Göttingen, I got acquainted with many great people, who have contributed in their own way to this work and to the person I have become. It remains a difficult task to acknowledge all of these people with concise wording, and therefore, I consider my attempt below also far from complete.

First and foremost, I thank my supervisor Aljaž Godec. I can say with great confidence that I could not have gotten a better supervisor. His office door has always been open for a discussion, a coffee, or a good Chuck Norris fact. Without his guidance and constant desire for a deeper understanding, I would not have been able to produce the work presented in this thesis. The way Aljaž thinks in terms of inequalities and counterexamples has been of great influence on my way of thinking. I also thank Aljaž's family who has shown great hospitality to me.

Secondly, I would like to thank my colleagues—those in the past and present—in the Mathematical bioPhysics Group. Alessio Lapolla and David Hartich were the first two colleagues I met, and I thank you both for your kindness and inspiring work attitude. Alessio convinced me to join his sports wrestling courses, which were a lot of fun, despite his superior strength. With David and his family I share many warm memories, and the visit to Esslingen was a highlight that hopefully will soon take place again. Together with Maximilian Vossel, I started my doctorate. Being the barista in our group, he has made many delicious espressos, and I thank you for your thoughtfulness and for the great moments we share. I thank Cai Dieball for his sharp questions, fruitful discussions, dry humor, and the rounds of tennis we played during the summer. With Janik Schüttler, I had many enjoyable moments on the running track, where he often pushed me to do an extra round. I thank you for your friendliness and positive spirit, hopefully we can once meet again in Cambridge. Finally, I thank Rick Bebon and Gerrit Wellecke for being such pleasant colleagues, who are always open to a discussion. Rick has extensively read every chapter of my thesis and provided me with great feedback. All in all, I am thankful for my colleagues who made/make the Mathematical bioPhysics Group a great place to work.

Within the *Theoretical and Computational Biophysics* department, there were/are also many colleagues to thank. Special thanks go to Lars Bock, Sara Gabrielli, Gabor

Nagy, Florian Leidner, Wojciech Kopec, and Peleg Sapir who have been thoughtful, friendly, and open to me. It was a pleasure to have lunch with you, meet for leisure time, or to have a little chat in the office. I thank Ansgar Esztermann and Martin Fechner for teaching me the ins and outs of computer cluster usage, which I initially did not use correctly. Furthermore, I thank Eveline Heinemann, Stefanie Teichmann, and Petra Kellers for the administrative assistance and their fast support during times I was rather late with my conference applications. Finally, I would like to thank Helmut Grubmüller and Bert de Groot for providing critical and honest feedback on my preparation talks. More than once, Helmut happily surprised me with a scientific riddle or an unexplained magic trick, which led to some interesting discussions and explanations.

As my Ph.D. thesis is done within the *International Max Planck Research School for Physics of Biological and Complex Systems*, I am indebted to Antje Erdmann and Frauke Bergmann for their administrative and practical support. Together we had various meetings which I experienced as very pleasant and successful.

I thank the members of my thesis advisory committee (TAC) Matthias Krüger and Peter Sollich for their valuable input during the TAC meetings and statistical physics seminars we had throughout the years. Furthermore I thank our collaborators David Zwicker and Noah Ziethen with whom we had various pleasant and insightful meetings.

Since my first day in Göttingen, I am living together with the caring family Trümper. A warm home is of invaluable significance for the well-being of anyone, and therefore, I thank Lorenz, Elisabeth, and Lukas for offering me a lovely and safe place to live.

It goes without saying that I also thank my amazing family. My parents Edwin and Lettie Blom have always supported my decisions and made it possible for me to pursue my study in physics and biology at the Technical University of Delft. The visits of my family to Göttingen I will not quickly forget, and hopefully, there are many more to come. I am truly thankful for all of you.

Lastly, I want to thank Mareile Steinsiek and her family. I met Mareile in the second year of my Ph.D. while following a Lindy Hop (dancing) course. Since then we have been together, and she always cheered me up no matter how discouraged I was. Mareile has shown a great interest in my work, which often brought my enthusiasm back in times of struggle. Thank you for that Mareilie. I thank Peter and Maja Steinsiek for the cozy and delightful visits we had with them. I am looking forward to many more visits to come, you are truly wonderful people.

Contents

Acronyms

AFM	Atomic force microscopy
BG	Bethe–Guggenheim
CAM	Cellular adhesion molecule
CVA	Cluster variation approximation
DB	Detailed balance
E-L	Euler–Lagrange
EOS	Equation of state
FPT	First passage time
HJ	Hamilton–Jacobi
LEQ	Local equilibrium approximation
LHS	Left-hand side
MC	Monte Carlo
MF	Mean field
MFPT	Mean first passage time
ODE	Ordinary differential equation
PDE	Partial differential equation
RHS	Right-hand side
SM	Supplementary material
SOS	Solid-on-solid

Chapter 1
Introduction

Starting around 1925, a change occurred: With the work of Ising, statistical mechanics began to be used to describe the behavior of many particles at once.

Kadanoff [1]

From the smallest subatomic scales to the largest extragalactic scales, it is eminent that the physical principles governing the dynamics of a single isolated particle are inadequate to understand the world around us. Interacting many-body systems are rather the rule than the exception, and therefore a large part of statistical physics is devoted to this topic. Of all interacting many-body systems in statistical physics on which exact calculations have been conducted, the **Ising model** is by far the most thoroughly studied [2], with every year around 700–900 published on the subject [3]. However, despite the vast amount of exact results in one- and two-dimensions, there is still uncovered ground where exact calculations have not reached. In this Chapter we present a brief history of the Ising model, show its application to biological systems, and reveal the questions addressed in this thesis.

1.1 The Ising Model: A Brief History

Covering the full history of the Ising model would be worth a Ph.D. thesis on its own. Here we highlight the most prominent results which are important for this thesis. To be precise, any *detailed* discussion related to critical exponents, magnetic frustration, the Hubbard-Stratonovich transformation, relations to conformal field theory, the renormalization group, Yang-Lee zeros, or dimensions equal or higher than three, are omitted from this section.

K. Blom, *Pair-Correlation Effects in Many-Body Systems*, Springer Theses,
https://doi.org/10.1007/978-3-031-29612-3_1

1.1.1 Spontaneous Magnetization in One Dimension (1920)

The Ising model was originally developed in 1920 by the German Wilhelm Lenz [4] and later analyzed by his student Ernst Ising[1] [6] to explain the mechanism behind **ferromagnetism**.[2] Ferromagnetism is the phenomenon by which certain materials spontaneously become magnetic when the temperature drops below a critical value, known as the **Curie temperature** [7]. An example is iron, which is magnetic below 1043 K Kelvin [8]. Lenz and Ising were inspired by the work of Pierre Weiss who assumed that every atom in a ferromagnetic material possesses a microscopic magnetic dipole σ_i, also called a "*spin*", which can point in one of two directions: up ($\sigma_i = +1$) or down ($\sigma_i = -1$) [9]. These spins are located on a lattice and interact with their immediate neighbors through a **coupling strength** J. Furthermore, Lenz and Ising introduced an external magnetic field h which points in the up ($h > 0$) or down ($h < 0$) direction. The well-known Hamiltonian of the Ising model with N spins $\{\boldsymbol{\sigma}\} = \{\sigma_1, \sigma_2, ..., \sigma_N\}$ reads

$$\mathcal{H}(\{\boldsymbol{\sigma}\}) = -J \sum_{\langle ij \rangle} \sigma_i \sigma_j - h \sum_i \sigma_i, \tag{1.1}$$

where the sum $\langle ij \rangle$ goes over nearest neighbor interactions. For a **ferromagnetic** coupling with $J > 0$, two aligned nearest-neighbor spins with $\sigma_i \sigma_j = +1$ decrease the energy. Similarly, two anti-aligned nearest-neighbor spins with $\sigma_i \sigma_j = -1$ increase the energy. Therefore, a ferromagnetic coupling J drives spins to an aligned configuration where the energy is lowest. For an **antiferromagnetic** coupling with $J < 0$ we have the opposite scenario, and the ground state is set by the staggered configuration where each spin is anti-aligned with its neighbor. Being in contact with an equilibrium heat bath at finite temperature T, the spins undergo random state fluctuations. The equilibrium probability for a configuration $\{\boldsymbol{\sigma}\}$ is given by $p(\{\boldsymbol{\sigma}\}) = \exp(-\beta \mathcal{H}(\{\boldsymbol{\sigma}\}))/\mathcal{Z}$, where $\beta = 1/k_{\mathrm{B}}T$ with k_{B} the Boltzmann constant, and $\mathcal{Z}$ is the **canonical partition function** given by a sum over all possible 2^N states

$$\mathcal{Z} = \sum_{\{\boldsymbol{\sigma}\}} \mathrm{e}^{-\beta \mathcal{H}(\{\boldsymbol{\sigma}\})}. \tag{1.2}$$

In his celebrated work, Ising computed the partition function for a one-dimensional chain composed of N spins analytically [6]. Furthermore, he computed analytically the average magnetization per spin $M(h) \in [-1, 1]$, which is defined as

$$M(h) \equiv (N\mathcal{Z})^{-1} \sum_{\{\boldsymbol{\sigma}\}} \mathrm{e}^{-\beta \mathcal{H}(\{\boldsymbol{\sigma}\})} \sum_{i=1}^{N} \sigma_i. \tag{1.3}$$

[1] Ernst Ising studied Mathematics and Physics in Göttingen in [5].

[2] Comes from the word ferro + magnetic: behaving like iron in a magnetic field.

Since Eq. (1.1) is invariant under the transformation $(h, \sigma_i) \to -(h, \sigma_i)$, it follows directly that $M(h) = -M(h)$, which implies that $M(0) = 0$. However, upon taking the thermodynamic limit, $M(h)$ is not guaranteed to be analytic at $h = 0$ [2]. The non-analyticity of $M(h)$ follows from the infinite number of terms inside the sum of Eq. (1.3). **Spontaneous magnetization** is therefore defined as

$$\lim_{h\to 0^\pm} \lim_{N\to\infty} \pm M(h) > 0. \tag{1.4}$$

Ising found no spontaneous magnetization at non-zero temperatures for the one-dimensional chain and concluded [6]: "... *wir finden also keine Hysteresiserscheinungen und keine spontane Magnetisierung.*" (p. 256). Hence, for the one-dimensional Ising model there is no spontaneous magnetization at non-zero temperature in the thermodynamic limit. But Ising went further, and erroneously concluded, in the same paper, that there is no spontaneous magnetization in two dimensions or higher: "*Auch bei einem räumlichen Modell, das den oben aufgestellten Annahmen genügt, gelangt man nicht zu einem anderen Ergebnis.*" (p. 257). This statement baffled the physics community, and started a longstanding scientific pursuit for spontaneous magnetization in two dimensions.

1.1.2 The Bragg-Williams Approximation (1934)

In 1934 an approximate analysis for the critical temperature in two dimensions was developed by Sir William Lawrence Bragg and Evan James Williams [10]. Their work focused on binary alloys,[3] but there is a close analogy with the Ising model as alluded by themselves [10]: "*The general conclusion that order sets in abruptly below a critical temperature T_c has a close analogy in ferromagnetism, ...*" (p. 707). Interest in binary alloys came forth from the hypothesis of Gustav Tamann, who stated that atoms in an alloy segregate under annealing[4] [11]. Considering two types of atoms A and B with fixed concentrations and a repulsive interaction, Bragg and Williams calculated the probability to interchange an A–B pair with an A–A or B–B pair. To do so, they assumed that the concentration of pairs is given by the product of respective A and B concentrations. This is also known as the zeroth-order or mean field **MF** approximation. The temperature where the mixed state with equal A and B concentration becomes unstable matches the critical temperature. Below this temperature a transition occurs, where the A and B atoms segregate into macroscopic clusters. Most surprisingly, Bragg and Williams found a critical temperature in two dimensions. Unfortunately, their approximation also predicted the existence of a critical temperature in one dimension, contradicting the exact results by Ising, and rendering their conclusion for two dimensions untrustworthy.

---

[3] Binary alloys are metals composed of two species. An example is cupronickel = cupper + nickel.

[4] Annealing refers to a heat treatment which reduces the stiffness of metals.

1.1.3 The Bethe-Guggenheim Approximation (1935)

Hans Bethe [12] and Edward Armand Guggenheim [13] independently published an improvement of the Bragg-Williams approximation in 1935. Interestingly, Guggenheim did not work on binary alloys, but on the theory of liquids with short-range interactions. The main improvement of their method was that the concentration of A-B pairs is not *exclusively* determined by the product of A and B concentrations, but *in addition* by the work required to convert an A-A and B-B pair into two A-B pairs. This improvement eradicated the fallacious critical temperature in one dimension. Moreover, their approximation provided the first evidence for spontaneous magnetization in the two-dimensional Ising model, as pointed out by Sir Rudolf Ernst Peierls [14]: "*Although the good convergence of Bethe's method is reassuring, it is worth while to give a rigorous proof for the ferromagnetic behaviour at low temperatures.*" (p. 478). At present, the **Bethe-Guggenheim (BG)** approximation is also known as the Bethe-Peierls or quasi-chemical approximation, based on the work of Peierls [15] and Guggenheim [16]. Nonetheless, in compliance with the historical order, we will refer to the BG approximation. In this thesis the BG approximation will be used in various settings, and in Chap. 2 we provide a more detailed explanation.

1.1.4 Spontaneous Magnetization in Two Dimensions (1936)

In 1936, twelve years after the erroneous conclusion made by Ising, Peierls proved that for sufficiently low temperatures the two-dimensional Ising model on a square lattice has spontaneous magnetization [14].[5] However, the exact critical temperature where $M(h)$ becomes non-analytic was unknown. Five years later, in 1941, Hendrik Anthony Kramers and Gregory Hugh Wannier introduced the transfer matrix method and showed that the critical temperature was given by $\sinh(2\beta J) = 1$[18]. Subsequently, in a second publication, they obtained highly accurate approximate closed-form expressions for the partition function per spin $\mathcal{Z}^{1/N}$ with zero field $h = 0$ [19]. Lars Onsager was inspired by the transfer matrix method [20], and considered the anisotropic two-dimensional Ising model with ferromagnetic couplings (J_x, J_y) and zero field $h = 0$. In 1943 he published his seminal work [21], with *exact* closed-form expressions for the partition function per spin and critical temperature $\sinh(2\beta J_x)\sinh(2\beta J_y) = 1$, generalizing the result of Kramers and Wannier. The spontaneous magnetization remained undetermined, until Onsager unveiled his result five years later at a conference in Cornell [20]:

$$\lim_{h\to 0^\pm} \lim_{N\to\infty} \pm M(h) = \left(1 - \sinh^{-2}(2\beta J_x)\sinh^{-2}(2\beta J_y)\right)^{1/8}. \tag{1.5}$$

[5] The original proof by Peierls turned out to be incorrect and was later refined in [17].

A derivation was never published by Onsager nor by his collaborator Bruria Kaufman with whom he had correspondence [22, 23]. Eventually, Chen Ning Yang re-derived and verified his result in [24].[6] Of honourable mention is the derivation of the partition function using dimer coverings, introduced by Mark Kac and John Clive Ward in [26].

1.1.5 Intermezzo: The Fixed-Magnetization Ensemble

Considering its long standing history, one might be tempted to think that most properties of the two-dimensional Ising model are currently known. However, the BG approximation provides information where the exact methods developed by Kramer, Wannier, and Onsager cannot reach. Furthermore, the BG approximation becomes exact on the Bethe lattice [27]. Recall that in the MF and BG approximation the concentration of A and B atoms was fixed. Identifying the distinct atoms with a spin up and down state, this is equivalent to fixing the magnetization in the system. Let $N_{\downarrow}(\{\boldsymbol{\sigma}\}) \in [0, N]$ be the number of down spins for a given configuration $\boldsymbol{\sigma}$. As will be detailed in Chap. 2, both the MF and BG approximation provide a procedure to evaluate the partition function in the **fixed-magnetization ensemble**, given by

$$\mathcal{Z}_k = \sum_{\{\sigma\}} \mathrm{e}^{-\beta\mathcal{H}(\{\sigma\})} \delta_{N_{\downarrow}(\{\sigma\}),k}, \tag{1.6}$$

where $\delta_{a,b}$ is Kronecker's delta function, i.e. $\delta_{a,b} = 1$ when $a = b$ and 0 elsewhere. The partition function $\mathcal{Z}_k$ contains all states with k down spins, and therefore a fixed magnetization per spin of $M = 1 - 2k/N$. Obtaining $\mathcal{Z}_k$ through exact methods would be a pièce de résistance, as it automatically gives an expression for the canonical partition function *with* an external magnetic field ($h \neq 0$) through the relation $\mathcal{Z} = \sum_{k=0}^{N} \mathcal{Z}_k$. For the two-dimensional square lattice Ising model this remains an unsolved problem [28–32], and exact expressions have *only* been derived for a triangular lattice [33], Kagome lattice [34], and the 3–12 lattice [35]. Thus, whereas $\mathcal{Z}$ is well-known for the two-dimensional square lattice, $\mathcal{Z}_k$ is not, and the BG approximation provides an accurate method to derive the latter. In the Chaps. 2–6 we shall see that $\mathcal{Z}_k$ plays a dominant role in many of the questions we will tackle in this thesis.

1.1.6 The Lattice Gas (1952)

First introduced by Tsung-Dao Lee and Cheng Ning Yang in [36], the lattice gas is a model where each lattice site can either be empty ($b_i = 0$) or occupied ($b_i = 1$) by

[6] According to Yang "*the longest calculation in my career*" (p. 371) in [25].

an atom. For an attractive interaction between nearest-neighbor atoms, there exists a critical temperature below which the atoms condense into a macroscopic phase, also known as a **phase transition**. Lee and Yang showed that the lattice gas and Ising model are isomorphic, by relating the number of atoms in the lattice gas to the number of down (or up) spins in the Ising model.[7] The Grand partition function of the lattice gas is therefore mathematically identical to the partition function of the Ising model in the presence of a magnetic field. Similarly, the partition function in the lattice gas corresponds to the fixed-magnetization partition function in the Ising model. This translation allowed Lee and Yang to study the nature of phase transitions through the exact results obtained by Onsager.

1.1.7 Kikuchi's Cluster Variation Approximation (1952)

The methods established by Bragg, Williams, Bethe, and Guggenheim culminated in a systematic approach for the fixed-magnetization partition function, developed by Ryoichi Kikuchi in [37]. The so-called **cluster variation approximation (CVA)** cleared the way for more accurate evaluations of the critical temperature for *any kind of lattice including higher dimensions*, and has been extremely successful ever since (Google Scholar: 2383 citations at 02/05/2023). The CVA establishes an approximation to the partition sum in Eq. 1.6 by approximating the number of distinct lattice configurations at a fixed energy; i.e. the **degeneracy of states**. By counting the filling statistics of pre-defined spin clusters in the lattice, one can approximate the degeneracy of states in a controlled way. The larger the clusters, the more accurate the approximation. With the resulting degeneracy factor, the partition sum in Eq. 1.6 is evaluated through a variational approach. Notably, the CVA also applies to multicomponent systems, and can be extended to long-range interactions [38].

1.1.8 Cahn-Hilliard Free Energy for Non-uniform Systems (1958)

Until 1958, the spatial organization of up and down spins in the Ising model was unknown territory. This changed when John Cahn and John Hilliard introduced a free energy formalism for non-uniform systems, providing insight into the spatial structure of non-uniform binary mixtures[8] [39]. If we ought to be historically precise, we should emphasize that Johannes Diderik van der Waals introduced an identical formalism in [40].[9] Unfortunately, due to the First World War, van der Waals did not receive much credit for his work [42]: "*Like most of his work, it was well-*

[7] This can be established through the mapping $\sigma_i = 2b_i - 1$.

[8] Binary mixtures are mixtures composed of two components, as for example the Ising model.

[9] This work is in Dutch. An English translation is provided in [41].

received at the time, but grievously neglected in the forty years after 1914." (p. 198). With the work of Cahn, Hilliard, and van der Waals, it became possible to study **concentration profiles**, **surface tension**, **spinodal decomposition**, and **nucleation phenomena**, within one general framework. Although a connection with the Ising model is not mentioned in [39], we will show in Chap. 3 that this formalism can be obtained *directly* from the partition function of the Ising model. Going beyond the MF approximation employed in [39], we will derive the free energy functional of a non-uniform Ising model which explicitly takes into account pair correlations. The resulting (and surprising) outcomes of this theory are further discussed in Chap. 4.

1.1.9 Stochastic Dynamics of the Ising Model (1963)

Where Cahn, Hilliard, and van der Waals introduced a spatial analysis of the Ising model, Roy Jay Glauber introduced a temporal analysis in [43]: "*We have attempted, therefore, to devise a form of the Ising model whose behavior can be followed exactly, in statistical terms, as a function of time.*" (p. 294). Considering **single-spin flip** dynamics, Glauber constructed a continuous-time **master equation** with the transition rates uniquely specified by symmetry, and the **detailed balance (DB)** relation.[10] Solving the master equation for a one-dimensional ring, he obtained the average magnetization, spin-spin correlation function, and time-delayed correlation function, for finite and infinite systems. Additionally, he considered a time-varying external magnetic field, and provided the first fluctuation-dissipation theorem[11] for the Ising model.

Three years later, Kyozi Kawasaki established another approach to study time-dependent behavior of the Ising model [44]. Instead of single-spin flip dynamics, Kawasaki considered **two-spin exchange** dynamics where two spins can interchange their position, hence conserving the total magnetization. Similar to Glauber, Kawasaki constructed a continuous-time master equation with a unique expression for the transition rates. Consecutively, he determined the diffusion coefficient near the critical temperature, where it converges to zero. This is known as critical slowing down, as the dynamics slows down and comes to a halt at the critical temperature.

The work of Glauber and Kawasaki opened the door for kinetic studies on the Ising model, and in Chaps. 5 and 6 we will study their formalism in more detail. Present-day, Glauber and Kawasaki dynamics on the coarse-grained continuum level also go under the name Model A and Model B, respectively, based on the classification made by Bert Halperin and Pierre Claude Hohenberg [45]. In Fig. 1.1 some biological examples of Glauber and Kawasaki dynamics are given.

---

[10] DB is a sufficient condition for the rates to approach equilibrium in the long time limit.

[11] This is a relation between the response function and equilibrium correlation function.

1.1.10 Intermezzo: Continuous and Discrete-Time Dynamics

In Chap. 5 we will encounter the **discrete-time** variant of Glauber and Kawasaki dynamics, whereas in Chap. 6 we consider **continuous-time** dynamics. Both types of dynamics are related through their **generators**, which evolve the corresponding probability vectors $\mathbf{P}_d(n),\ n \in \mathbb{N}$ (discrete-time) and $\mathbf{P}_c(t),\ \forall t \geq 0$ (continuous-time). To see this, let $\mathbf{Q}_d$ and $\mathbf{Q}_c$ be the respective (finite-dimensional) generators of the discrete and continuous-time Markov process, such that $\mathbf{P}_d(n+1) = \mathbf{Q}_d\mathbf{P}_d(n)$ and $\partial_t\mathbf{P}_c(t) = \mathbf{Q}_c\mathbf{P}_c(t)$. For the Ising model, the generators are two-dimensional matrices of size $2^N \times 2^N$, where N is the number of spins. Taking the equilibrium steady-state $\mathbf{P}^{\text{ss}}_{d,c}$, whose existence is guaranteed by DB [46], we obtain $(\mathbf{Q}_d - \mathbb{1})\mathbf{P}^{\text{ss}}_d = 0$ and $\mathbf{Q}_c\mathbf{P}^{\text{ss}}_c = 0$, where $\mathbb{1}$ is the identity matrix. Enforcing equal equilibrium steady-states for continuous and discrete time dynamics, i.e. $\mathbf{P}^{\text{ss}}_c = \mathbf{P}^{\text{ss}}_d$, we obtain the following non-unique solution for the generators $\mathbf{Q}_d - \mathbb{1} = \mathbf{Q}_c$.[12] Hence, if the generators for discrete and continuous-time dynamics are related by $\mathbf{Q}_d - \mathbb{1} = \mathbf{Q}_c$, then they have the same equilibrium steady state [46].

1.1.11 Kadanoff's Block Spin Method (1966)

In 1966 Leo Kadanoff devised a scheme to obtain a relation between critical exponents[13] in the Ising model, through the so-called **block spin method** [47]. The basic idea behind the block spin method is to group spins together, and coarse-grain them to a new variable, the so-called block spin. The partition function attains a new form under the coarse-grained block spins. However, the old form can effectively be restored upon introducing a new effective interaction strength J'. The relation between the new effective interaction strength J' and old interaction strength J induces a flow, the so-called renormalization flow. Although we are not concerned with critical exponents and the renormalization group, in Chap. 3 we will apply a similar block spin technique to derive a Cahn-Hilliard free energy functional from the fixed-magnetization partition function of the Ising model.

1.1.12 The Ising Model at Present Day (2000–2022)

In modern-day physics the Ising model remains to play an active role. Ranging from high-energy physics [48, 49] to econophysics [50, 51], it is clear that the model has entered a diverging set of disciplines. Recent works discovered a **finite-time dynamical phase transition** in the MF **Curie-Weiss** model [52–55]. The dynamical phase transition appears for **disordering quenches** in the Ising model, whereby the

---

[12] Note that $\mathbf{Q}_d - \mathbb{1} = \mathbf{Q}_c \Rightarrow (\mathbf{Q}_d - \mathbb{1})\mathbf{P}^{\text{ss}} = \mathbf{Q}_c\mathbf{P}^{\text{ss}}$, but $(\mathbf{Q}_d - \mathbb{1})\mathbf{P}^{\text{ss}} = \mathbf{Q}_c\mathbf{P}^{\text{ss}} \not\Rightarrow \mathbf{Q}_d - \mathbb{1} = \mathbf{Q}_c$.

[13] Critical exponents are universal scaling factors of observables around a critical point.

coupling J or temperature T is abruptly changed to a new value $J \to J' < J$ or $T \to T' > T$, which induces a temporal evolution of the spins towards a state with less ferromagnetic order. In Chap. 6 we will analyze disordering quenches within the BG approximation, and find that the finite-time dynamical phase transition is also present in the nearest-neighbor Ising model.

1.2 Biological Applications of the Ising Model

In the field of (molecular) biology the Ising model has found many applications. The list of examples is extensive, and include: bacterial chemotaxis proteins [56, 65, 66], **cell adhesion** [59, 67–75], chemical sensing [76–78], biomolecular condensates [79–82], cancer [83–85], allosteric proteins [86–88], cooperative ligand binding [89–91], cooperative gating in ion channels [62–64, 92], chromosome folding [93, 94], DNA folding [95–98], protein folding [99], microtubules [100–102], genomics [103, 104], neuroscience [105–108], epidemic modeling [109–111], and even phenotypic traits of reptiles [60]. In most of these applications there exist a cooperative degree of freedom which can attain two states, i.e. the presence and absence of a hydrogen bond in a DNA molecule. These degrees of freedom are written as Ising spins ($\sigma_i = \pm 1$) or lattice gas variables ($b_i = 0 \vee 1$), which provides a map to the Ising model. A subset of examples is depicted in Fig. 1.1. It should be emphasized that not every application of the Ising model is well-justified, and some appear to be rather unorthodox.[14] However, the abundance of examples underlines the widespread usage of the model outside the field of physics. Below, we provide a short introduction of one specific application relevant for Chap. 5; cell adhesion.

1.2.1 *Cell Adhesion*

An adult human body consists of approximately 3×10^{13} eukaryotic cells, of which 10% belong to solid tissue. These cells are physically attached through a process known as cell adhesion. Cell adhesion refers to the specific binding of cells to neighboring cells or the extracellular matrix, and plays a major role in cell regulation [113], intercellular communication [114], immune response [115], wound healing [116], morphogenesis [117], cellular function [118], and tumorigenesis [119, 120].

Cellular adhesion domains form as a result of the association of transmembrane cellular adhesion molecule (CAM)s that are interacting with the actin cytoskeleton [121] and can translocate over the membrane [122]. There are four major superfamilies of CAMs: the immunoglobulins, integrins, cadherins, and selectins. Throughout we generically refer to them as CAMs. Adhesion bonds are typically non-covalent, with binding energies on the order of a few $k_{\mathrm{B}}T$ corresponding to forces on the

[14] An example here would be the application to love [112].

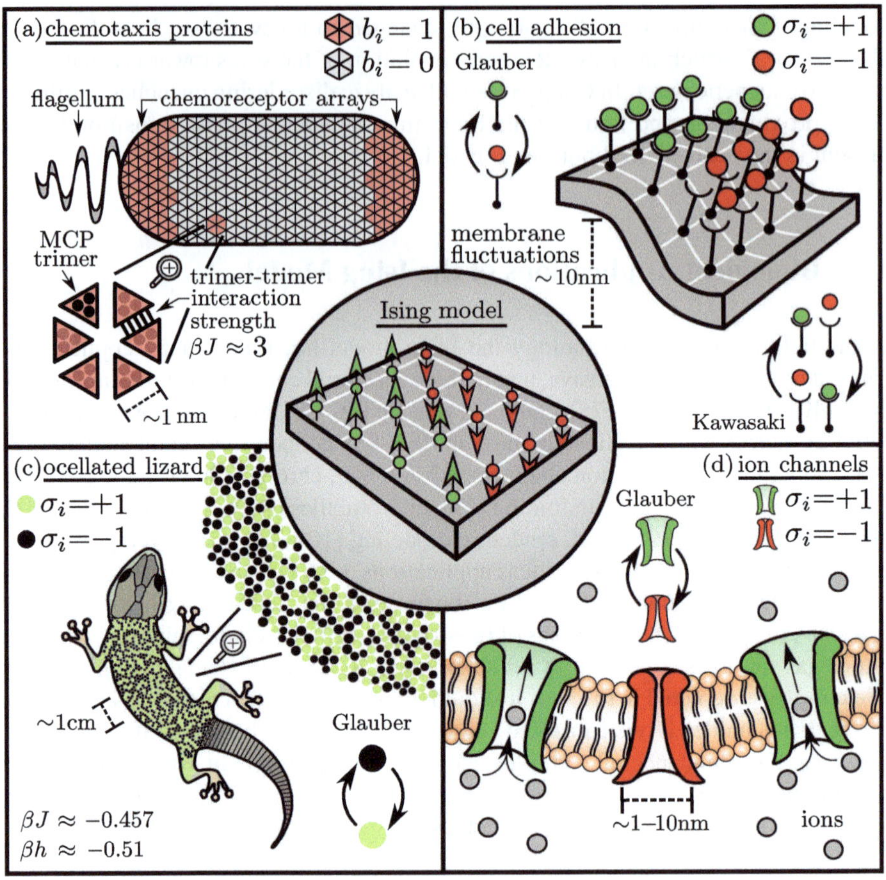

Fig. 1.1 Biological applications of the Ising model. a Hexagonally ordered methyl-accepting chemotaxis proteins (MCP) cluster into trimers and localize at bacterial cell poles, forming a highly structured hexagonal array. In [56] trimers are modeled as discrete lattice gas variables (based on [57]) with attractive trimer-trimer coupling $\beta J \approx 3$. **b** Cellular adhesion molecules coupled to a fluctuating membrane. The nanometric membrane fluctuations mediate a lateral interaction between nearest-neighbor adhesion states [58], which can be mapped onto an Ising model [59]. **c** Ocellated lizard *(Timon lepidus)* with black and green monochromatic skin scales. The scale patterns on the back of the reptile are well-captured by an Ising model with antiferromagnetic coupling $\beta J \approx -0.457$ and magnetic field $\beta h \approx -0.51$ [60]. **d** Ion channels are pore-forming membrane proteins enabling transmembrane ion transport. Cooperative gating refers to a coupling of gating states between neighbouring ion channels, which has been observed in various experiments [61]. In [62–64] the Ising model with open ($\sigma_i = 1$) and closed ($\sigma_i = -1$) gating states is utilized to model the effect of cooperative gating

order of $\simeq 4\,\text{pN} \cdot \text{nm}$ at $T \simeq 300\,\text{K}$ [123, 124]. As a result of thermal fluctuations, these bonds have finite lifetimes – they can break and re-associate depending on the receptor-ligand distance, their respective conformations and local concentrations, and depending on internal and external mechanical forces [124, 125].

While it was originally thought that adhesion strength is determined by the biochemistry of adhesion molecules alone, more recently, cellular mechanics [126] and membrane-induced adhesion bond interactions [59, 72, 127–129] emerged as essential physical regulators of cellular adhesion. In fact, there exists an exact mapping between cell adhesion and the lattice gas, where the coupling strength J is set by the membrane properties and binding affinities of the receptors [59, 71–73] (see also Fig. 1.1b). In Chap. 5 we will employ this mapping between cell adhesion and the Ising model, to study the equilibrium and kinetic properties of adhesion clusters in the absence and presence of an external force.

1.3 Scope and Outline of the Thesis

Having introduced the paradigmatic Ising model, its relevance in cellular biological systems, and the notion of the MF and BG approximation, we now lay forth the structure of the forthcoming chapters:

- In Chaps. 2 and 3 we introduce the technical background needed for the remaining Chaps. 4–6. Here we employ the BG approximation to incorporate nearest neighbor pair correlations into the fixed-magnetization partition function of the Ising model. In Chap. 2 we apply the BG approximation to uniform systems, and in Chap. 3 to non-uniform systems. The latter results in a Cahn-Hilliard free energy functional with a **concentration dependent gradient energy coefficient**.
- In Chap. 4 we systematically analyze the resulting outcomes of Chap. 3. Here we find that the Cahn-Hilliard free energy functional obtained with the BG approximation takes into account the effect of capillary waves, giving rise to a **delocalization-induced interface widening**.
- Chap. 5 focuses on the role of cooperativity in cell adhesion. Using the mapping to the lattice gas provided by [59], we construct an equilibrium and kinetic description of cell adhesion under the presence of an external force. Focusing on the dynamics of cluster formation/dissolution, we show the emergence of a **dynamical critical coupling** in the thermodynamic limit, where formation/dissolution times are the fastest.
- In Chap. 6 we use the BG approximation to analyze the non-equilibrium relaxation of the Ising model under disordering quenches. Here we demonstrate the existence of a **finite-time dynamical phase transition** in the nearest-neighbor Ising model, and furthermore show that the appearance of the phase transition obeys a speed limit.
- Finally, in Chap. 7 we conclude and provide an outlook into further extensions and possible outcomes.

References

1. Kadanoff LP (2013) Relating theories via renormalization. Stud Hist Philos Sci B 44(1):22–39 (2013). [Online]. Available https://www.sciencedirect.com/science/article/pii/S1355219812000391
2. McCoy BM, Wu TT (2014) The two-dimensional Ising model. Courier Corporation
3. Kobe S (1997) Ernst Ising—physicist and teacher. J Stat Phys 88(3):991–995. [Online]. Available https://doi.org/10.1023/B:JOSS.0000015184.19421.03
4. Lenz W (1920) Beitrag zum verständnis der magnetischen erscheinungen in festen körpern. Z Phys 21:613–615. [Online]. Available https://www.physik.uni-rostock.de/storages/uni-rostock/Alle_MNF/Physik/Historisches/Kalenderblaetter_Physik/KB_2013_06_Lenz/Lenz_1920.pdf
5. Brush SG (1967) History of the Lenz-Ising model. Rev Mod Phys 39:883–893. [Online]. Available https://link.aps.org/doi/10.1103/RevModPhys.39.883
6. Ising E (1925) Beitrag zur Theorie des Ferromagnetismus. Zeitschrift für Physik 31(1):253–258. [Online]. Available http://link.springer.com/10.1007/BF02980577
7. Curie P (1895) Propriétés magnétiques des corps a diverses températures. Ann Chim Phys 5(289). [Online]. Available https://iris.univ-lille.fr/handle/1908/466
8. Laughlin DE (2018) The β iron controversy revisited. J Phase Equilibria Diffus 39(3):274–279. [Online]. Available https://doi.org/10.1007/s11669-018-0638-z
9. Weiss P (1907) L'hypothèse du champ moléculaire et la propriété ferromagnétique. J Phys Theor Appl 6(1):661–690. [Online]. Available https://doi.org/10.1051/jphystap:019070060066100
10. Bragg WL, Williams EJ (1934) The effect of thermal agitation on atomic arrangement in alloys. Proc Math Phys Eng Sci 145(855):699–730. [Online]. Available https://royalsocietypublishing.org/doi/abs/10.1098/rspa.1934.0132
11. Tammann, G (1919) Zum gedächtnis der entdeckung des isomorphismus vor 100 jahren. die chemischen und galvanischen eigenschaften von mischkristallreihen und ihre atomverteilung. Z Anorg Allg Chem 107(1):1–239. [Online]. Available https://onlinelibrary.wiley.com/doi/abs/10.1002/zaac.19191070102
12. Bethe HA (1935) Statistical theory of superlattices. Proc Math Phys Eng Sci 150(871):552–575. [Online]. Available https://royalsocietypublishing.org/doi/abs/10.1098/rspa.1935.0122
13. Guggenheim EA (1935) The statistical mechanics of regular solutions. Proc Math Phys Eng Sci 148(864):304–312. [Online]. Available https://royalsocietypublishing.org/doi/abs/10.1098/rspa.1935.0020
14. Peierls R (1936) On Ising's model of ferromagnetism. Math Proc Camb Philos Soc 32(3):477–481
15. Peierls R, Bragg WL (1936) Statistical theory of superlattices with unequal concentrations of the components. Proc R Soc Lond Ser A Math Phys Sci 154(881):207–222. [Online]. Available https://royalsocietypublishing.org/doi/abs/10.1098/rspa.1936.0047
16. Fowler RH, Guggenheim EA (1940) Statistical thermodynamics of super-lattices. Proc R Soc Lond Ser A Math Phys Sci 174(957):189–206. [Online]. Available https://royalsocietypublishing.org/doi/abs/10.1098/rspa.1940.0014
17. Griffiths RB (1964) Peierls proof of spontaneous magnetization in a two-dimensional Ising ferromagnet. Phys Rev 136:A437–A439. [Online]. Available https://link.aps.org/doi/10.1103/PhysRev.136.A437
18. Kramers HA, Wannier GH (1941) Statistics of the two-dimensional ferromagnet. Part I. Phys Rev 60:252–262. [Online]. Available https://link.aps.org/doi/10.1103/PhysRev.60.252
19. Kramers HA, Wannier GH (1941) Statistics of the two-dimensional ferromagnet. Part II. Phys Rev 60:263–276. [Online]. Available https://link.aps.org/doi/10.1103/PhysRev.60.263
20. Longuet-Higgins HC, Fisher ME (1978) Lars Onsager, 27 November–5 October 1976. Biogr Mem Fellows R Soc 24:443–471. [Online]. Available https://royalsocietypublishing.org/doi/abs/10.1098/rsbm.1978.0014

21. Onsager L (1944) Crystal statistics. I. A two-dimensional model with an order-disorder transition. Phys Rev 65:117–149. [Online]. Available https://link.aps.org/doi/10.1103/PhysRev.65.117
22. Baxter RJ (2011) Onsager and Kaufman's calculation of the spontaneous magnetization of the Ising model. J Stat Phys 145:518–548. [Online]. Available https://doi.org/10.1007/s10955-011-0213-z
23. Baxter RJ (2012) Onsager and Kaufman's calculation of the spontaneous magnetization of the Ising model II. J Stat Phys 149:1164–1167. [Online]. Available https://doi.org/10.1007/s10955-012-0658-8
24. Yang CN (1952) The spontaneous magnetization of a two-dimensional Ising model. Phys Rev 85:808–816. [Online]. Available https://link.aps.org/doi/10.1103/PhysRev.85.808
25. Fisher ME (2015) Statistical physics in the oeuvre of Chen Ning Yang. Int J Mod Phys B 29(30):1530013. [Online]. Available https://doi.org/10.1142/S0217979215300133
26. Kac M, Ward JC (1952) A combinatorial solution of the two-dimensional Ising model. Phys Rev 88:1332–1337. [Online]. Available https://link.aps.org/doi/10.1103/PhysRev.88.1332
27. Domb C (1960) On the theory of cooperative phenomena in crystals. Adv Phys 9(34):149–244. [Online]. Available https://doi.org/10.1080/00018736000101189
28. Stošć B, Milošević S, Stanley HE (1990) Exact results for the two-dimensional Ising model in a magnetic field: tests of finite-size scaling theory. Phys Rev B 41:11466–11478. [Online]. Available https://link.aps.org/doi/10.1103/PhysRevB.41.11466
29. Zamolodchikov A (1989) Integrable field theory from conformal field theory. In: Jimbo M, Miwa T, Tsuchiya A (eds) Integrable system quantum field theory. Academic Press, San Diego, pp 641–674. [Online]. Available https://www.sciencedirect.com/science/article/pii/B9780123853424500226
30. Fonseca P (2003) Ising field theory in a magnetic field: analytic properties of the free energy. J Stat Phys 110:527–590. [Online]. Available https://doi.org/10.1023/A:1022147532606
31. Mangazeev VV, Dudalev MY, Bazhanov VV et al (2010) Scaling and universality in the two-dimensional Ising model with a magnetic field. Phys Rev E 81:060103. [Online]. Available https://link.aps.org/doi/10.1103/PhysRevE.81.060103
32. McCoy BM, Maillard J-M (2012) The importance of the Ising model. Prog Theor Phys 127(5):791–817. [Online]. Available https://doi.org/10.1143/PTP.127.791
33. Wu FY (1985) Exact solution of a triangular Ising model in a nonzero magnetic field. J Stat Phys 40(5):613–620. [Online]. Available https://doi.org/10.1007/BF01009892
34. Giacomini H (1988) An exact result for the Kagome lattice Ising model with magnetic field. J Phys A Math Theor 21(1):L31–L36. [Online]. Available https://doi.org/10.1088/0305-4470/21/1/006
35. Lin K (1989) Exact results for the Ising model on a 3–12 lattice. Int J Mod Phys B 03(08):1237–1245. [Online]. Available https://doi.org/10.1142/S021797928900083X
36. Lee TD, Yang CN (1952) Statistical theory of equations of state and phase transitions. II. Lattice gas and Ising model. Phys Rev 87:410–419. [Online]. Available https://link.aps.org/doi/10.1103/PhysRev.87.410
37. Kikuchi R (1951) A theory of cooperative phenomena. Phys Rev 81:988–1003. [Online]. Available https://link.aps.org/doi/10.1103/PhysRev.81.988
38. Morita T (1957) Cluster variation method of cooperative phenomena and its generalization I. J Phys Soc Jpn 12(7):753–755. [Online]. Available https://doi.org/10.1143/JPSJ.12.753
39. Cahn JW, Hilliard JE (1958) Free energy of a nonuniform system. I. Interfacial free energy. J Chem Phys 28(2):258–267. [Online]. Available https://aip.scitation.org/doi/10.1063/1.1744102
40. Van der Waals J (1893) Thermodynamische theorie der capillariteit in de onderstelling van continue dichtheidsverandering, verhand. Kon Akad Wetensch Amsterdam Sect 1
41. van der Waals JD (1979) The thermodynamic theory of capillarity under the hypothesis of a continuous variation of density. J Stat Phys 20(2):200–244. [Online]. Available https://doi.org/10.1007/BF01011514

42. Rowlinson JS (1979) Translation of J. D. van der Waals "the thermodynamik theory of capillarity under the hypothesis of a continuous variation of density". J Stat Phys 20(2):197–200. [Online]. Available https://doi.org/10.1007/BF01011513
43. Glauber RJ (1963) Time-dependent statistics of the Ising model. J Math Phys 4(2):294–307. [Online]. Available https://aip.scitation.org/doi/10.1063/1.1703954
44. Kawasaki K (1966) Diffusion constants near the critical point for time-dependent Ising models. I. Phys Rev 145:224–230. [Online]. Available https://link.aps.org/doi/10.1103/PhysRev.145.224
45. Hohenberg PC, Halperin BI (1977) Theory of dynamic critical phenomena. Rev Mod Phys 49:435–479. [Online]. Available https://link.aps.org/doi/10.1103/RevModPhys.49.435
46. Norris JR (1998) Markov chains, no 2. Cambridge University Press. Cambridge, UK
47. Kadanoff LP (1966) Scaling laws for Ising models near T_c. Phys Physique Fizika 2:263–272. [Online]. Available https://link.aps.org/doi/10.1103/PhysicsPhysiqueFizika.2.263
48. Iqbal N, McGreevy J (2020) Toward a 3D Ising model with a weakly-coupled string theory dual. SciPost Phys 9:19. [Online]. Available https://scipost.org/10.21468/SciPostPhys.9.2.019
49. Sulejmanpasic T (2021) Ising model as a u(1) lattice gauge theory with a θ-term. Phys Rev D 103:034512. [Online]. Available https://link.aps.org/doi/10.1103/PhysRevD.103.034512
50. Schinckus C (2018) Ising model, econophysics and analogies. Physica A 508:95–103. [Online]. Available https://www.sciencedirect.com/science/article/pii/S0378437118306083
51. Pickhardt M, Seibold G (2014) Income tax evasion dynamics: evidence from an agent-based econophysics model. J Econ Psychol 40:147–160. [Online]. Available https://www.sciencedirect.com/science/article/pii/S0167487013000408
52. Meibohm J, Esposito M (2022) Finite-time dynamical phase transition in nonequilibrium relaxation. Phys Rev Lett 128:110603. [Online]. Available https://link.aps.org/doi/10.1103/PhysRevLett.128.110603
53. Meibohm J, Esposito M (2022) Landau theory for finite-time dynamical phase transitions. arXiv:2205.10311. [Online]. Available https://arxiv.org/abs/2205.10311
54. Külske C, Le Ny A (2007) Spin-flip dynamics of the Curie-Weiss model: loss of Gibbsianness with possibly broken symmetry. Commun Math Phys 271(2):431–454. [Online]. Available http://dx.doi.org/10.1007/s00220-007-0201-y
55. Ermolaev V, Külske C (2010) Low-temperature dynamics of the Curie-Weiss model: periodic orbits, multiple histories, and loss of Gibbsianness. J Stat Phys 141(5):727–756. [Online]. Available http://dx.doi.org/10.1007/s10955-010-0074-x
56. Endres RG (2009) Polar chemoreceptor clustering by coupled trimers of dimers. Biophys J 96(2):453–463. [Online]. Available https://doi.org/10.1016/j.bpj.2008.10.021
57. Mukhopadhyay R, Huang KC, Wingreen NS (2008) Lipid localization in bacterial cells through curvature-mediated microphase separation. Biophys J 95(3):1034–1049. [Online]. Available https://www.sciencedirect.com/science/article/pii/S0006349508701747
58. Fenz SF, Bihr T, Schmidt D et al (2017) Membrane fluctuations mediate lateral interaction between cadherin bonds. Nat Phys 13(9):906–913. [Online]. Available https://www.nature.com/articles/nphys4138
59. Speck T, Reister E, Seifert U (2010) Specific adhesion of membranes: mapping to an effective bond lattice gas. Phys Rev E 82:021923. [Online]. Available https://link.aps.org/doi/10.1103/PhysRevE.82.021923
60. Zakany S, Smirnov S, Milinkovitch MC (2022) Lizard skin patterns and the Ising model. Phys Rev Lett 128:048102. [Online]. Available https://link.aps.org/doi/10.1103/PhysRevLett.128.048102
61. Dixon RE, Navedo MF, Binder MD et al (2022) Mechanisms and physiological implications of cooperative gating of clustered ion channels. Physiol Rev 102(3):1159–1210. pMID: 34927454. [Online]. Available https://doi.org/10.1152/physrev.00022.2021
62. Liu Y, Dilger JP (1993) Application of the one-and two-dimensional Ising models to studies of cooperativity between ion channels. Biophys J 64(1):26–35

63. Erdem R (2006) Collective equilibrium behaviour of ion channel gating in cell membranes: an Ising model formulation. J Biol Phys 32(6):523–529. [Online]. Available https://doi.org/10.1007/s10867-007-9034-3
64. Erdem R, Aydiner E (2009) Monte Carlo simulation for statistical mechanics model of ion-channel cooperativity in cell membranes. Phys Rev E 79:031919. [Online]. Available https://link.aps.org/doi/10.1103/PhysRevE.79.031919
65. Shimizu TS, Aksenov SV, Bray D (2003) A spatially extended stochastic model of the bacterial chemotaxis signalling pathway. J Mol Biol 329(2):291–309. [Online]. Available https://www.sciencedirect.com/science/article/pii/S0022283603004376
66. Mello BA, Shaw L, Tu Y (2004) Effects of receptor interaction in bacterial chemotaxis. Biophys J 87(3):1578–1595. [Online]. Available https://doi.org/10.1529/biophysj.104.042739
67. Kraus M, Wolf B (1994) Ising model for cooperative processing of extracellular information by protein-tyrosine kinases and cell adhesion molecules. Biosystems 33(2):89–98. [Online]. Available https://www.sciencedirect.com/science/article/pii/0303264794900493
68. Khain E, Sander LM, Schneider-Mizell CM (2007) The role of cell-cell adhesion in wound healing. J Stat Phys 128(1):209–218. [Online]. Available https://doi.org/10.1007/s10955-006-9194-8
69. Blom K, Godec A (2021) Criticality in cell adhesion. Phys Rev X 11(3):031067. [Online]. Available https://link.aps.org/doi/10.1103/PhysRevX.11.031067
70. Zhao J, Wu J, Veatch S (2013) Adhesion stabilizes robust lipid heterogeneity in supercritical membranes at physiological temperature. Biophys J 104(4):825–834. [Online]. Available https://www.sciencedirect.com/science/article/pii/S0006349513000350
71. Speck T, Vink RLC (2012) Random pinning limits the size of membrane adhesion domains. Phys Rev E 86:031923. [Online]. Available https://link.aps.org/doi/10.1103/PhysRevE.86.031923
72. Speck T (2011) Effective free energy for pinned membranes. Phys Rev E 83:050901. [Online]. Available https://link.aps.org/doi/10.1103/PhysRevE.83.050901
73. Speck T (2011) Publisher's note: Effective free energy for pinned membranes [phys. rev. e 83, 050901(r) (2011)]. Phys Rev E 83:059909. [Online]. Available https://link.aps.org/doi/10.1103/PhysRevE.83.059909
74. Albert PJ, Schwarz US (2016) Modeling cell shape and dynamics on micropatterns. Cell Adh Migr 10(5):516–528. pMID: 26838278. [Online]. Available https://doi.org/10.1080/19336918.2016.1148864
75. Albert P, Schwarz U (2014) Dynamics of cell shape and forces on micropatterned substrates predicted by a cellular Potts model. Biophys J 106(11):2340–2352. [Online]. Available https://www.sciencedirect.com/science/article/pii/S0006349514004548
76. Hu B, Chen W, Rappel W-J et al (2010) Physical limits on cellular sensing of spatial gradients. Phys Rev Lett 105:048104. [Online]. Available https://link.aps.org/doi/10.1103/PhysRevLett.105.048104
77. Shi Y, Duke T (1998) Cooperative model of bacterial sensing. Phys Rev E 58:6399–6406. [Online]. Available https://link.aps.org/doi/10.1103/PhysRevE.58.6399
78. Skoge M, Naqvi S, Meir Y et al (2013) Chemical sensing by nonequilibrium cooperative receptors. Phys Rev Lett 110:248102. [Online]. Available https://link.aps.org/doi/10.1103/PhysRevLett.110.248102
79. Weiner BG, Pyo AG, Meir Y et al (2021) Motif-pattern dependence of biomolecular phase separation driven by specific interactions. PLoS Comput Biol 17(12):e1009748. [Online]. Available https://doi.org/10.1371/journal.pcbi.1009748
80. Morin JA, Wittmann S, Choubey S et al (2022) Sequence-dependent surface condensation of a pioneer transcription factor on DNA. Nat Phys 18(3):271–276. [Online]. Available https://doi.org/10.1038/s41567-021-01462-2
81. Weber CA, Zwicker D, Jülicher F et al (2019) Physics of active emulsions. Rep Prog Phys 82(6):064601. [Online]. Available https://doi.org/10.1088/1361-6633/ab052b
82. Perdikari TM, Jovic N, Dignon GL et al (2021) A predictive coarse-grained model for position-specific effects of post-translational modifications. Biophys J 120(7):1187–1197. [Online]. Available https://www.sciencedirect.com/science/article/pii/S0006349521001351

83. Torquato S (2011) Toward an Ising model of cancer and beyond. Phys Biol 8(1):015017. [Online]. Available https://doi.org/10.1088/1478-3975/8/1/015017
84. Barradas-Bautista D, Alvarado-Mentado M, Agostino M et al (2018) Cancer growth and metastasis as a metaphor of go gaming: an Ising model approach. PLoS One 13(5):e0195654. [Online]. Available https://doi.org/10.1371/journal.pone.0195654
85. Züleyha A, Ziya M, Selçuk Y et al (2017) Simulation of glioblastoma multiforme (GBM) tumor cells using Ising model on the Creutz cellular automaton. Physica A 486:901–907. [Online]. Available https://www.sciencedirect.com/science/article/pii/S0378437117306155
86. Graham I, Duke TAJ (2005) Dynamic hysteresis in a one-dimensional Ising model: application to allosteric proteins. Phys Rev E 71:061923. [Online]. Available https://link.aps.org/doi/10.1103/PhysRevE.71.061923
87. LeVine MV, Weinstein H (2015) Aim for Allostery: using the Ising model to understand information processing and transmission in allosteric biomolecular systems. Entropy 17(5):2895–2918. [Online]. Available https://www.mdpi.com/1099-4300/17/5/2895
88. Bray D (2013) The propagation of allosteric states in large multiprotein complexes. J M Biol 425(9):1410–1414; allosteric interactions and biological regulation (part I). [Online]. Available https://www.sciencedirect.com/science/article/pii/S0022283612009412
89. Schwarz G (1970) Cooperative binding to linear biopolymers. Euro J Mol Biol Biochem 12(3):442–453. [Online]. Available https://febs.onlinelibrary.wiley.com/doi/abs/10.1111/j.1432-1033.1970.tb00871.x
90. Schmitz KS, Schurr JM (1970) Cooperative binding of adenosine by polyuridylic acid: a further analysis. Biopolymers 9(6):697–715. [Online]. Available https://onlinelibrary.wiley.com/doi/abs/10.1002/bip.1970.360090608
91. Stirmanov YV, Matveeva OV, Nechipurenko YD (2019) Two-dimensional Ising model for microarray hybridization: cooperative interactions between bound target molecules. J Biomol Struct Dyn 37(12):3103–3108. pMID: 30081753. [Online]. Available https://doi.org/10.1080/07391102.2018.1508370
92. Matsson L, Sa-Yakanit V, Boribarn S (2005) Lyotropic ion channel current model compared with Ising model. J Biol Phys 31(3):525–532. [Online]. Available https://doi.org/10.1007/s10867-005-4807-z
93. Colì D, Orlandini E, Michieletto D et al (2019) Magnetic polymer models for epigenetics-driven chromosome folding. Phys Rev E 100:052410. [Online]. Available https://link.aps.org/doi/10.1103/PhysRevE.100.052410
94. Colliva A, Pellegrini R, Testori A et al (2015) Ising-model description of long-range correlations in DNA sequences. Phys Rev E 91:052703. [Online]. Available https://link.aps.org/doi/10.1103/PhysRevE.91.052703
95. Vtyurina NN, Dulin D, Docter MW et al (2016) Hysteresis in DNA compaction by DPS is described by an Ising model. Proc Natl Acad Sci USA 113(18):4982–4987. [Online]. Available https://www.pnas.org/doi/abs/10.1073/pnas.1521241113
96. Liebl K, Zacharias M (2021) Accurate modeling of DNA conformational flexibility by a multivariate Ising model. Proc Natl Acad Sci USA 118(15):e2021263118. [Online]. Available https://www.pnas.org/doi/abs/10.1073/pnas.2021263118
97. Bloomfield VA (1999) Statistical thermodynamics of helix–coil transitions in biopolymers. Am J Phys 67(12):1212–1215. [Online]. Available https://doi.org/10.1119/1.19107
98. Mel'nikov SM, Sergeyev VG, Mel'nikova YS et al (1997) Folding of long DNA chains in the presence of distearyldimethylammonium bromide and unfolding induced by neutral liposomes. J Chem Soc Faraday Trans 93:283–288. [Online]. Available http://dx.doi.org/10.1039/A603753D
99. Makarov DE (2009) A theoretical model for the mechanical unfolding of repeat proteins. Biophys J 96(6):2160–2167. [Online]. Available http://www.sciencedirect.com/science/article/pii/S0006349509003208
100. Marchi E, Benegas J, Ripoll D (1978) Temperature dependence of microtubule structure. J Chem Phys 69(2):800–805. [Online]. Available https://doi.org/10.1063/1.436592

101. Trpišová B, Brown JA (1998) Ordering of dipoles in different types of microtubule lattice. Int J Mod Phys B 12(05):543–578. [Online]. Available https://doi.org/10.1142/S0217979298000338
102. Ripoll D, Marchi E, Benegas J (1981) A new approach to microtubule depolymerization. J Theor Biol 93(1):53–62. [Online]. Available https://www.sciencedirect.com/science/article/pii/0022519381900564
103. Majewski J, Li H, Ott J (2001) The Ising model in physics and statistical genetics. Am J Hum Gen 69(4):853–862. [Online]. Available https://doi.org/10.1086/323419
104. Fischer S, Wegener I (2004) The Ising model on the ring: mutation versus recombination. In: Deb K (ed) Genetic and evolutionary computation—GECCO 2004. Springer, Berlin, Heidelberg, pp 1113–1124. [Online]. Available https://link.springer.com/chapter/10.1007/978-3-540-24854-5_109
105. Hertz JA, Roudi Y, Thorning A et al (2010) Inferring network connectivity using kinetic Ising models. BMC Neurosci 11:51. [Online]. Available https://doi.org/10.1186/1471-2202-11-S1-P51
106. Marinazzo D, Pellicoro M, Wu G-R et al (2013) Information transfer of an Ising model on a brain network. BMC Neurosci 14(1):1–2. [Online]. Available https://doi.org/10.1186/1471-2202-14-S1-P376
107. Schaub MT, Schultz SR (2012) The Ising decoder: reading out the activity of large neural ensembles. J Comp Neurosci 32(1):101–118. [Online]. Available https://doi.org/10.1007/s10827-011-0342-z
108. Fraiman D, Balenzuela P, Foss J et al (2009) Ising-like dynamics in large-scale functional brain networks. Phys Rev E 79:061922. [Online]. Available https://link.aps.org/doi/10.1103/PhysRevE.79.061922
109. Serra N, Di Carlo P, Rea T et al (2021) Diffusion modeling of covid-19 under lockdown. Phys Fluids 33(4):041903. [Online]. Available https://doi.org/10.1063/5.0044061
110. Mello IF, Squillante L, Gomes GO et al (2021) Epidemics, the Ising-model and percolation theory: a comprehensive review focused on covid-19. Phys A 573:125963. [Online]. Available https://www.sciencedirect.com/science/article/pii/S0378437121002351
111. Machado D, Mulet R (2021) From random point processes to hierarchical cavity master equations for stochastic dynamics of disordered systems in random graphs: Ising models and epidemics. Phys Rev E 104:054303. [Online]. Available https://link.aps.org/doi/10.1103/PhysRevE.104.054303
112. Solnyshkov D, Malpuech G (2022) Love might be a second-order phase transition. Phys lett A 445:128245. [Online]. Available https://www.sciencedirect.com/science/article/pii/S0375960122003279
113. Huang S, Ingber DE (1999) The structural and mechanical complexity of cell-growth control. Nat Cell Biol 1(5):E131–E138. [Online]. Available http://dx.doi.org/10.1038/13043
114. Gille J, Swerlick RA (1996) Integrins: role in cell adhesion and communication. Ann N Y Acad Sci 797(1):93–106. [Online]. Available http://dx.doi.org/10.1111/j.1749-6632.1996.tb52952.x
115. Zarnitsyna V, Zhu C (2012) T cell triggering: insights from 2D kinetics analysis of molecular interactions. Phys Biol 9(4):045005. [Online]. Available https://doi.org/10.1088%2F1478-3975%2F9%2F4%2F045005
116. Brugués A, Anon E, Conte V et al (2014) Forces driving epithelial wound healing. Nat Phys 10(9):683–690. [Online]. Available http://dx.doi.org/10.1038/nphys3040
117. Borghi N, Nelson WJ (2009) Chapter 1 intercellular adhesion in morphogenesis: molecular and biophysical considerations. In: Current topics in developmental biology, series current topics in developmental biology, vol 89. Academic Press. [Online]. Available http://www.sciencedirect.com/science/article/pii/S0070215309890017
118. Hynes RO (1999) Cell adhesion: old and new questions. Trends Cell Biol 9(12):M33–37. [Online]. Available https://doi.org/10.1016/S0962-8924(99)01667-0
119. Cavallaro U, Christofori G (2004) Cell adhesion and signalling by cadherins and IG-cams in cancer. Nat Rev Cancer 4(2):118–132. [Online]. Available http://dx.doi.org/10.1038/nrc1276

120. Andl CD (2010) The misregulation of cell adhesion components during tumorigenesis: overview and commentary. J Oncol 2010. [Online]. Available https://doi.org/10.1155/2010/174715
121. Morone N, Nakada C, Umemura Y et al (2008) Chapter 12 three-dimensional molecular architecture of the plasma-membrane-associated cytoskeleton as reconstructed by freeze-etch electron tomography. In: Introduction to electron microscopy for biologists, series. Methods in cell biology, vol 88. Academic Press, pp 207 – 236. [Online]. Available http://www.sciencedirect.com/science/article/pii/S0091679X08004123
122. DeMond AL, Mossman KD, Starr T et al (2008) T cell receptor microcluster transport through molecular mazes reveals mechanism of translocation. Biophys J 94(8):3286–3292. [Online]. Available https://www.sciencedirect.com/science/article/pii/S0006349508704855
123. Bell G (1978) Models for the specific adhesion of cells to cells. Science 200(4342):618–627. [Online]. Available http://www.sciencemag.org/cgi/doi/10.1126/science.347575
124. Schwarz US, Safran SA (2013) Physics of adherent cells. Rev Mod Phys 85(3):1327–1381. [Online]. Available https://link.aps.org/doi/10.1103/RevModPhys.85.1327
125. Evans E, Ritchie K (1997) Dynamic strength of molecular adhesion bonds. Biophys J 72(4):1541–1555. [Online]. Available https://linkinghub.elsevier.com/retrieve/pii/S0006349597788027
126. Schmitz J, Gottschalk K-E (2008) Mechanical regulation of cell adhesion. Soft Matter 4(7):1373. [Online]. Available http://dx.doi.org/10.1039/B716805P
127. Zuckerman D, Bruinsma R (1995) Statistical mechanics of membrane adhesion by reversible molecular bonds. Phys Rev Lett 74:3900–3903. [Online]. Available https://link.aps.org/doi/10.1103/PhysRevLett.74.3900
128. Bihr T, Seifert U, Smith A-S (2012) Nucleation of ligand-receptor domains in membrane adhesion. Phys Rev Lett 109(25):258101. [Online]. Available https://link.aps.org/doi/10.1103/PhysRevLett.109.258101
129. Farago O (2010) Fluctuation-induced attraction between adhesion sites of supported membranes. Phys Rev E 81(5 Pt 1):050902

Chapter 2
Bethe-Guggenheim Approximation for Uniform Systems

Obsession with partition functions maketh a dull man.

Lars Onsager to Joseph Hubbard [1] (p. 457)

In this chapter we apply the BG approximation to uniform lattice systems. We consider an Ising model defined on a general lattice with isotropic coupling J, and calculate the **fixed-magnetization partition function**. Subsequently, we take the thermodynamic limit and calculate the **free energy density**, **phase diagram**, and **free energy minima**. Finally, we provide a comparison between the BG and MF results and show that qualitative differences emerge in the strong coupling limit. For other accounts of the BG approximation, see e.g. [2–4]. Part of this chapter has been published in *Physical Review X* under the terms of the Creative Commons Attribution 4.0 International license [5].

2.1 Initial Setup and Definitions

We consider an Ising model on a general lattice with average lattice coordination number $\bar{z}$. Examples are the one-dimensional line ($\bar{z} = 2$), two-dimensional square lattice ($\bar{z} = 4$), hexagonal lattice ($\bar{z} = 3$), and triangular lattice ($\bar{z} = 6$). Note that different lattices can have equal average coordination numbers. For example: the triangular, simple cubic, and Bethe lattice with six neighbors all have $\bar{z} = 6$ as shown in Fig. 2.1. The MF and BG approximation do not distinguish between these lattices, and only for the Bethe lattice it provides an exact result. Let φ denote the **fraction of down spins**. The **thermodynamic limit** is defined as the scaling limit where we take the number of spins N to infinity, while simultaneously fixing the fraction of down spins, $\varphi = k/N$, i.e.

K. Blom, *Pair-Correlation Effects in Many-Body Systems*, Springer Theses,
https://doi.org/10.1007/978-3-031-29612-3_2

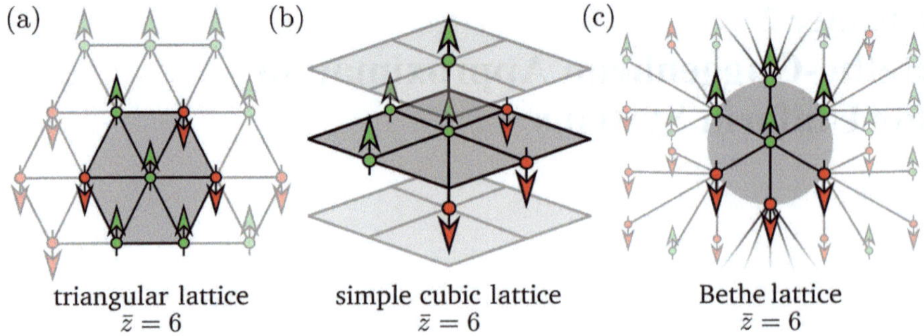

Fig. 2.1 **Example of different lattices with equal average coordination number $\bar{\mathbf{z}}$**. The **a** triangular, **b** simple cubic, and **c** Bethe lattice with six neighbors all have $\bar{z} = 6$

$$\lim_{\mathrm{s}}^{N} [\cdot] \equiv \lim_{\varphi=k/N=\mathrm{const.}}^{N\to\infty} [\cdot] \,. \tag{2.1}$$

The fraction of down spins $\varphi \in [0, 1]$ becomes a continuous variable in the thermodynamic limit. The free energy density in units of thermal energy $k_{\mathrm{B}}T$ (i.e. $\tilde{\mathrm{f}} \equiv \mathrm{f}/k_{\mathrm{B}}T$) for a given fraction of down spins, is given by

$$\tilde{\mathrm{f}}(\varphi = k/N) \equiv \lim_{\mathrm{s}}^{N}[-N^{-1} \ln{(\mathcal{Z}_k)}], \tag{2.2}$$

where $\mathcal{Z}_k$ is the fixed-magnetization partition function defined in Eq. (1.6). The free energy density can subsequently be written as $\tilde{\mathrm{f}}(\varphi) = \tilde{\mathrm{u}}(\varphi) - \tilde{\mathrm{s}}(\varphi)$, with $\tilde{\mathrm{u}}(\varphi)$ the **internal energy density** and $\tilde{\mathrm{s}}(\varphi)$ the **entropy density**. We now determine $\mathcal{Z}_k$ within the BG approximation, and after that we evaluate $\tilde{\mathrm{f}}(\varphi)$ via Eq. (2.2).

2.2 Rewriting the Ising Hamiltonian

Before we evaluate $\mathcal{Z}_k$, it will be useful to rewrite the Ising Hamiltonian. The magnetic field term coupled to h in Eq. (1.1) becomes constant upon fixing the magnetization, and can therefore be omitted from further analysis. The sum in the Ising Hamiltonian goes over nearest-neighbor terms, so we can rewrite Eq. (1.1) as

$$\mathcal{H}(\{\boldsymbol{\sigma}\}) = -J \sum_{\langle ij \rangle} \sigma_i \sigma_j = J(N_{\uparrow\downarrow} - N_{\uparrow\uparrow} - N_{\downarrow\downarrow}), \tag{2.3}$$

where $N_{\uparrow\uparrow}$, $N_{\downarrow\downarrow}$, and $N_{\uparrow\downarrow}$ denote the total number of up-up, down-down, and up-down spin pairs, respectively. Every down-down spin pair consists of two down spins, and every up-down spin pair consists of a single up and a single down spin; hence $2N_{\downarrow\downarrow} + N_{\uparrow\downarrow} = \bar{z}N_{\downarrow}$, where $N_{\downarrow}$ is the total number of down spins. This relation is

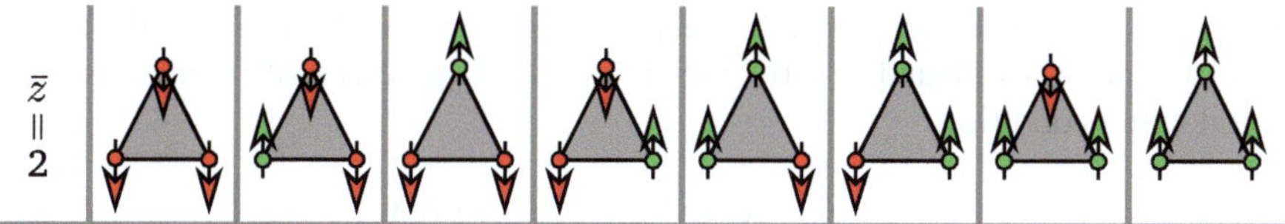

$N_{\downarrow}$	3	2	2	2	1	1	1	0
$N_{\uparrow\downarrow}$	0	2	2	2	2	2	2	0
$N_{\uparrow\uparrow}$	0	0	0	0	1	1	1	3
$N_{\downarrow\downarrow}$	3	1	1	1	0	0	0	0

Fig. 2.2 Example for the relations between $\mathbf{N_{\downarrow}, N_{\uparrow\downarrow}, N_{\uparrow\uparrow}, N_{\downarrow\downarrow}}$ given by Eq. (2.4)

exact for lattices with a constant coordination number per spin, e.g. for the Bethe lattice or a lattice with periodic boundary conditions. A similar reasoning applies to up-up spin pairs, resulting in the general relations

$$N_{\uparrow\uparrow} = (\bar{z}(N - N_{\downarrow}) - N_{\uparrow\downarrow})/2, \quad N_{\downarrow\downarrow} = (\bar{z}N_{\downarrow} - N_{\uparrow\downarrow})/2, \tag{2.4}$$

where we have used that $N = N_{\uparrow} + N_{\downarrow}$. In Fig. 2.2 we provide an example for three spins on a ring ($\bar{z} = 2$) where we can see that Eq. (2.4) is satisfied. Plugging Eq. (2.4) into Eq. (2.3) we obtain

$$\mathcal{H}(\{\sigma\}) = J(2N_{\uparrow\downarrow} - \bar{z}N/2), \tag{2.5}$$

Hence, we have re-written the Ising Hamiltonian in terms of the number of up-down spin pairs $N_{\uparrow\downarrow}$, also known as **defects**, **grain boundaries**, or the **short-range order** [6].

2.3 Rewriting the Partition Function

Plugging Eq. (2.5) into Eq. (1.6) we obtain an expression for the fixed-magnetization partition function in terms of defects. Instead of taking the partition sum over all configurations $\{\boldsymbol{\sigma}\}$ with $N_{\uparrow\downarrow}$ defects, we may formally sum over all distinct values of $N_{\uparrow\downarrow}$ and account for their multiplicity by introducing a **degeneracy of states** $\Psi(k, N_{\uparrow\downarrow})$ that counts the number of configurations with a given $N_{\uparrow\downarrow}$ at fixed k. This leads to the expression

$$\mathcal{Z}_k = \sum_{\{\sigma\}} \mathrm{e}^{\tilde{J}(\bar{z}N/2 - 2N_{\uparrow\downarrow})} \delta_{N_{\downarrow}(\{\sigma\}),k} = \sum_{N_{\uparrow\downarrow}} \Psi(k, N_{\uparrow\downarrow}) \mathrm{e}^{\tilde{J}(\bar{z}N/2 - 2N_{\uparrow\downarrow})}, \tag{2.6}$$

where $\tilde{J} \equiv \beta J$. The last expression in Eq. (2.6) remains exact, given that the degeneracy of states is evaluated exactly. For $\tilde{J} = 0$ the degeneracy of states must obey the following condition

$$\mathcal{Z}_k|_{\tilde{J}=0} = \sum_{N_{\uparrow\downarrow}} \Psi(k, N_{\uparrow\downarrow}) \stackrel{!}{=} \binom{N}{k}, \tag{2.7}$$

since this is the number of different configurations with k down spins. This condition will be used to construct an accurate approximation for the degeneracy of states, and thereby for $\mathcal{Z}_k$.

2.4 Pair Approximation Ansatz

We now employ the BG approximation to find a suitable expression for the degeneracy of states. Let us consider placing pairs of spins randomly onto a given lattice. Then, the total number of different lattice configurations for fixed $(N_{\uparrow\uparrow}, N_{\uparrow\downarrow}, N_{\downarrow\downarrow})$ may be approximated by

$$\Psi_{\text{BG}}(k, N_{\uparrow\downarrow}) \approx \mathcal{N}(k) \frac{(N_{\downarrow\downarrow} + N_{\uparrow\downarrow} + N_{\uparrow\uparrow})!}{(N_{\downarrow\downarrow})!\,(N_{\uparrow\downarrow}/2)!^2\,(N_{\uparrow\uparrow})!}, \tag{2.8}$$

where $\mathcal{N}(k)$ is a normalization constant to be determined by Eq. (2.7). For a general lattice, the number of defects can be odd, rendering $(N_{\uparrow\downarrow}/2)!$ ill-defined. Furthermore, it will be useful to introduce the **intensive variables** $(\varphi, \zeta) \equiv (k/N, N_{\uparrow\downarrow}/(\bar{z}N))$, which are non-integers. Therefore, we replace the factorial with the Gamma function (i.e. the generalized factorial), and substitute Eq. (2.4) for $(N_{\downarrow\downarrow}, N_{\uparrow\uparrow})$, which gives

$$\Psi_{\text{BG}}(\varphi, \zeta) \approx \frac{\mathcal{N}(\varphi)\hat{\Gamma}(1)}{\hat{\Gamma}(\varphi - \zeta)\,\hat{\Gamma}(\zeta)^2\,\hat{\Gamma}(1 - \varphi - \zeta)}, \tag{2.9}$$

where we introduced the function $\hat{\Gamma}(x) \equiv \Gamma(\bar{z}Nx/2 + 1)$ with $\Gamma(n) = (n-1)!$ for $n \in \mathbb{Z}^+$.

2.5 Evaluating the Normalization Constant

The normalization constant is determined by Eq. (2.7). To evaluate the sum over $N_{\uparrow\downarrow}(= \bar{z}N\zeta)$ we apply an analytic continuation of the **maximum term method** [7, 8] to real numbers. In other words, we approximate the sum by taking the largest term of the respective summand. To evaluate the largest term, we apply the thermodynamic limit and use **Stirling's approximation** for the Gamma function: $\ln \Gamma(w) = \Xi(w) - w + \mathcal{O}(\ln w)$ for $\mathrm{Re}(w) > 0$ with $\Xi(w) \equiv w \ln(w)$.[1] This gives the following result for the largest argument

$$\zeta_{\mathrm{BG}}(\varphi) \equiv \arg \sup_{\zeta}\{\lim_{\mathrm{s}}^{N}[N^{-1} \ln(\Psi_{\mathrm{BG}}(\varphi, \zeta))]\} = \varphi(1-\varphi). \tag{2.10}$$

This result is intuitive, as it agrees with the fraction of defects for a random mixture with a fraction of φ down spins. Evaluating the sum in Eq. (2.7) by the largest term, we obtain the following expression for the normalization constant

$$\mathcal{N}(\varphi) \stackrel{!}{=} \binom{N}{N\varphi} \Psi_{\mathrm{BG}}^{-1}(\varphi, \zeta_{\mathrm{BG}}(\varphi)). \tag{2.11}$$

This renders the degeneracy of states fully defined in terms of the variables (φ, ζ).

2.6 Evaluating the Partition Sum

Next we determine the partition sum over $N_{\uparrow\downarrow}(= \bar{z}N\zeta)$ in Eq. (2.6). Similarly as before, we apply the maximum term method, yielding the largest argument

$$\begin{aligned}\zeta_{\mathrm{BG}}^{\dagger}(\varphi) &\equiv \arg \sup_{\zeta}\{\lim_{\mathrm{s}}^{N}[N^{-1} \ln(\Psi(\varphi, \zeta)\mathrm{e}^{-2\bar{z}\tilde{J}N\zeta})]\} \\ &= \frac{2\zeta_{\mathrm{BG}}(\varphi)}{1+[1+4\zeta_{\mathrm{BG}}(\varphi)(\mathrm{e}^{4\tilde{J}}-1)]^{1/2}}.\end{aligned} \tag{2.12}$$

Two remarks related to $\zeta_{\mathrm{BG}}^{\dagger}(\varphi)$ are in order: First, it gives the thermodynamically optimal fraction of defects at a given coupling strength $\tilde{J}$ and fraction of down spins φ. For increasing $\tilde{J}$ the number of defects decreases, which is expected, since defects increase the energy. Second, for $\tilde{J} = 0$ we obtain $\zeta_{\mathrm{BG}}^{\dagger}(\varphi) = \zeta_{\mathrm{BG}}(\varphi)$, as expected. With the largest argument evaluated, the BG fixed-magnetization partition function becomes

$$\mathcal{Z}_{k=N\varphi}^{\mathrm{BG}} = \binom{N}{N\varphi} \frac{\Psi_{\mathrm{BG}}(\varphi, \zeta_{\mathrm{BG}}^{\dagger}(\varphi))}{\Psi_{\mathrm{BG}}(\varphi, \zeta_{\mathrm{BG}}(\varphi))} \mathrm{e}^{-2\bar{z}N\tilde{J}(\zeta_{\mathrm{BG}}^{\dagger}(\varphi)-1/4)}. \tag{2.13}$$

[1] We therefore expect the accuracy of the BG approximation to increase with N.

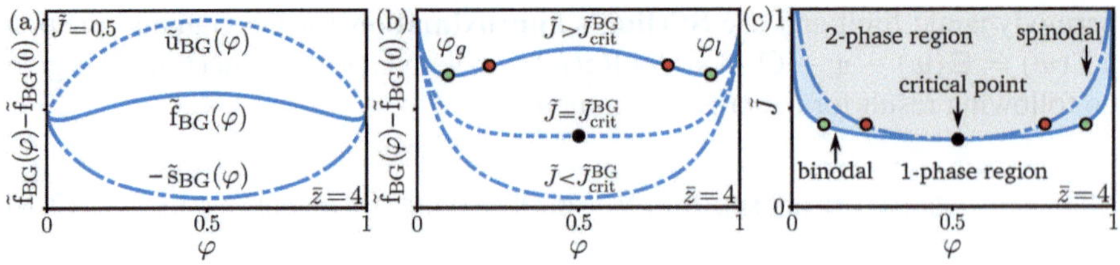

Fig. 2.3 **Free energy density and phase diagram of the BG approximation**. In each plot we consider a lattice with $\bar{z} = 4$ and vanishing external field $\tilde{h} = 0$. **a** Decomposition of the BG free energy density $\tilde{f}_{BG}(\varphi)$ (solid line) into the internal energy density $\tilde{u}_{BG}(\varphi)$ (dashed line) and entropy density $\tilde{s}_{BG}(\varphi)$ (dash-dotted line) given by Eq. (2.14). The coupling is fixed at $\tilde{J} = 0.5$. **b** Free energy density $\tilde{f}_{BG}(\varphi)$ for various values of $\tilde{J}$ below (dash-dotted line), at (dotted line), and above (solid line) the critical coupling $\tilde{J}^{BG}_{crit}$ given by Eq. (2.26). Above the critical coupling there exists two global minima, indicated with the green dots. The red dots indicate the inflection points where $\tilde{f}''_{BG}(\varphi) = 0$. **c** Phase diagram of the BG approximation in the $(\varphi, \tilde{J})$-plane. The binodal (solid line), spinodal (dash-dotted), and critical point (black dot) are given by Eqs. (2.20), (2.25), and (2.26), respectively. The gray area indicates the 2-phase region inside the spinodal where a uniform phase de-mixes into a liquid and gaseous phase through spinodal decomposition. The blue area indicates the region between the binodal and spinodal where uniform phases are metastable and de-mix through nucleation. The green and red dots located at the binodal and spinodal correspond to the green and red dots in **b**

2.7 Free Energy Density

Inserting Eq. (2.13) into Eq. (2.2) we determine the BG free energy density. Again we employ Stirling's approximation to evaluate the logarithm of the Gamma function. For completeness we also restore the magnetic field $\tilde{h} \equiv \beta h$, and obtain the following expressions for the internal energy and entropy density, i.e. $\tilde{f}_{BG}(\varphi) = \tilde{u}_{BG}(\varphi) - \tilde{s}_{BG}(\varphi)$, with

$$\begin{aligned}\tilde{u}_{BG}(\varphi) &= 2\bar{z}\tilde{J}[\zeta^{\dagger}_{BG}(\varphi) - 1/4] - \tilde{h}(1 - 2\varphi),\\ \tilde{s}_{BG}(\varphi) &= (\bar{z} - 1)[\Xi(\varphi) + \Xi(1 - \varphi)] - (\bar{z}/2)[\Xi(\varphi - \zeta^{\dagger}_{BG}(\varphi))\\ &\quad + \Xi(1 - \varphi - \zeta^{\dagger}_{BG}(\varphi)) + 2\Xi(\zeta^{\dagger}_{BG}(\varphi))],\end{aligned} \tag{2.14}$$

where $\Xi(\varphi) \equiv \varphi \ln \varphi$. Equations (2.13)–(2.14) are the main results of the BG approximation for a uniform system. In Fig. 2.3a the free energy density, internal energy density, and entropy density are plotted for a lattice with $\bar{z} = 4$ and $(\tilde{J}, \tilde{h}) = (0.5, 0)$. Additionally, in Fig. 2.3b the free energy density is shown for three increasing values of $\tilde{J}$. We see that above a critical value, which we denote as the **critical coupling**[2] $\tilde{J}^{BG}_{crit}$, the free energy density has two global minima. The φ values of the minima deviate from the center $\varphi = 0.5$, indicating that configurations with unequal up/down spin concentrations become thermodynamically favourable. This is the origin of spontaneous magnetization, since unequal up/down spin concentrations result

[2] Also known as the critical or Curie temperature, since one can write $\tilde{J}^{BG}_{crit} = \beta^{BG}_{crit} J$.

in a non-vanishing average magnetization. We now determine the location of the global minima (for $\tilde{h} = 0$) and critical coupling, which are part of the so-called phase diagram.

2.8 Phase Diagram

We calculate the phase diagram related to the BG free energy density in the $(\varphi, \tilde{J})$-plane. The phase diagram contains three parts: the **binodal**, the **spinodal**, and the **critical point**.

2.8.1 Binodal

The binodal $\tilde{J}_b(\varphi)$, also known as the **coexistence curve**, forms a line in the $(\varphi, \tilde{J})$-plane where two distinct phases with respective down-spin concentration φ_l and φ_g can coexist through **phase separation**. Here φ_l and φ_g denote the liquid (i.e. high concentration) and gas (i.e. low concentration) phase, respectively. The binodal applies when the total concentration of up and down spins is conserved,[3] and is determined by the common tangent construction,[4] which is the solution of the coupled equations

$$\tilde{\mathrm{f}}'(\varphi_l) = \tilde{\mathrm{f}}'(\varphi_g), \quad \frac{\tilde{\mathrm{f}}(\varphi_l) - \tilde{\mathrm{f}}(\varphi_g)}{\varphi_l - \varphi_g} = \tilde{\mathrm{f}}'(\varphi_l), \tag{2.15}$$

where $\mathrm{f}'(\varphi) \equiv d f(\varphi)/d\varphi$. The external field $\tilde{h}$ which couples linearly to φ trivially cancels out in Eq. (2.15). This is expected, since the magnetic field becomes a constant when the concentration of up and down spins is conserved. We therefore set $\tilde{h} = 0$ for further analysis. Due to the $\mathbb{Z}_2$ symmetry of the Ising model for $\tilde{h} = 0$, the free energy obeys $\tilde{\mathrm{f}}_{\mathrm{BG}}(\varphi) = \tilde{\mathrm{f}}_{\mathrm{BG}}(1 - \varphi),\ \forall \varphi \in [0, 1]$. The solutions to Eq. (2.15) therefore follow $\varphi_l = 1 - \varphi_g$, resulting in the much simpler equation $\tilde{\mathrm{f}}'_{\mathrm{BG}}(\varphi_{g,l}) = 0$. We solve this equation for $\zeta^{\dagger}_{\mathrm{BG}}(\varphi)$, and then use the inverse relation of Eq. (2.12) given by

$$\tilde{J} = \frac{1}{4} \ln \left(\frac{(\varphi - \zeta^{\dagger}_{\mathrm{BG}}(\varphi))(1 - \varphi - \zeta^{\dagger}_{\mathrm{BG}}(\varphi))}{\zeta^{\dagger}_{\mathrm{BG}}(\varphi)^2} \right), \tag{2.16}$$

to obtain the binodal. To evaluate the total derivative of $\tilde{\mathrm{f}}_{\mathrm{BG}}(\varphi)$ w.r.t. φ we apply the chain rule

$$\tilde{\mathrm{f}}'_{\mathrm{BG}}(\varphi) = \partial_\varphi \tilde{\mathrm{f}}_{\mathrm{BG}}(\varphi) + \partial_{\zeta^{\dagger}_{\mathrm{BG}}} \tilde{\mathrm{f}}_{\mathrm{BG}}(\varphi) \cdot \partial_\varphi \zeta^{\dagger}_{\mathrm{BG}}(\varphi). \tag{2.17}$$

[3] Hence the binodal applies to equilibrated Kawasaki dynamics.

[4] Also known as the Maxwell construction.

Since $\zeta^{\dagger}_{\mathrm{BG}}(\varphi)$ is derived from maximizing the partition summand, we have $\partial_{\zeta^{\dagger}_{\mathrm{BG}}} \tilde{\mathrm{f}}_{\mathrm{BG}}(\varphi) = 0$. Hence, the second term in Eq. (2.17) vanishes, while the first term yields

$$\partial_{\varphi}\tilde{\mathrm{f}}_{\mathrm{BG}}(\varphi) = (1-\bar{z})\ln\left(\frac{\varphi}{1-\varphi}\right) + \frac{\bar{z}}{2}\ln\left(\frac{\varphi - \zeta^{\dagger}_{\mathrm{BG}}(\varphi)}{1-\varphi-\zeta^{\dagger}_{\mathrm{BG}}(\varphi)}\right). \tag{2.18}$$

Setting this equal to zero, and introducing the auxiliary variables $\chi_{\varphi} \equiv \varphi/(1-\varphi)$ and $\alpha \equiv (\bar{z}-1)/\bar{z}$, we find the solution for $\zeta^{\dagger}_{\mathrm{BG}}(\varphi)$ at the binodal

$$\zeta^{\dagger}_{\mathrm{BG}}(\varphi)|_{\tilde{J}=\tilde{J}_b} = \frac{\chi_{\varphi}}{1+\chi_{\varphi}}\frac{1-\chi_{\varphi}^{2\alpha-1}}{1-\chi_{\varphi}^{2\alpha}}. \tag{2.19}$$

Plugging Eq. (2.19) into Eq. (2.16) finally yields the result for the binodal

$$\tilde{J}_b^{\mathrm{BG}}(\varphi) = \frac{1}{2}\ln\left(\frac{1-\chi_{\varphi}}{\chi_{\varphi}^{1/\bar{z}} - \chi_{\varphi}^{1-1/\bar{z}}}\right), \tag{2.20}$$

which is also reported in [2–4]. For $\bar{z} = 2$ the binodal diverges, since there is no phase separation in one dimension. In Fig. 2.3c the binodal for a lattice with $\bar{z} = 4$ is shown with the solid line.

2.8.2 Spinodal

The spinodal $\tilde{J}_s(\varphi)$, also known as the **stability boundary**, denotes the boundary between metastable and unstable uniform phases, and is determined by the inflection points $\tilde{\mathrm{f}}''(\varphi)=0$. In between the binodal and spinodal a uniform concentration is metastable, and can only undergo phase separation by **nucleation**. Within the spinodal a uniform concentration is unstable, and will undergo phase separation by **spinodal decomposition**. We calculate the second derivative of the BG free energy density, which reads

$$\tilde{\mathrm{f}}''_{\mathrm{BG}}(\varphi) = \frac{1-\bar{z}}{\varphi(1-\varphi)} + \frac{\bar{z}}{2}\frac{[1-2\zeta^{\dagger}_{\mathrm{BG}} + 2(2\varphi-1)(\zeta^{\dagger}_{\mathrm{BG}})']\zeta^{\dagger}_{\mathrm{BG}} + [2\varphi(1-\varphi) - \zeta^{\dagger}_{\mathrm{BG}}](\zeta^{\dagger}_{\mathrm{BG}})'^2}{\zeta^{\dagger}_{\mathrm{BG}}(\varphi - \zeta^{\dagger}_{\mathrm{BG}})(1-\varphi-\zeta^{\dagger}_{\mathrm{BG}})}, \tag{2.21}$$

where we suppressed the argument of $\zeta_{\mathrm{BG}}(\varphi)$ for convenience. Similarly to the binodal, we want to solve for the root of Eq. (2.21) in terms of $\zeta^{\dagger}_{\mathrm{BG}}(\varphi)$. Unfortunately, Eq. (2.21) contains $\zeta^{\dagger}_{\mathrm{BG}}(\varphi)'$, which we have to express in terms of $\zeta^{\dagger}_{\mathrm{BG}}(\varphi)$ and φ. Differentiating Eq. (2.16) w.r.t. φ at fixed $\tilde{J}$ gives

$$\zeta_{\mathrm{BG}}^{\dagger}(\varphi)' = \frac{\zeta_{\mathrm{BG}}^{\dagger}(\varphi)\,(1-2\varphi)}{2\varphi\,(1-\varphi) - \zeta_{\mathrm{BG}}^{\dagger}(\varphi)}. \tag{2.22}$$

Inserting Eq. (2.22) into Eq. (2.21) results in the much simpler expression

$$\tilde{\mathrm{f}}_{\mathrm{BG}}''(\varphi) = \frac{1-\bar{z}}{\varphi\,(1-\varphi)} + \frac{\bar{z}}{2\varphi\,(1-\varphi) - \zeta_{\mathrm{BG}}^{\dagger}(\varphi)}. \tag{2.23}$$

Setting Eq. (2.23) equal to zero and solving for $\zeta_{\mathrm{BG}}^{\dagger}(\varphi)$ yields

$$\zeta_{\mathrm{BG}}^{\dagger}(\varphi)|_{\tilde{J}=\tilde{J}_s} = (\bar{z}-2)\varphi(1-\varphi)/(\bar{z}-1). \tag{2.24}$$

Finally, we insert Eq. (2.24) into Eq. (2.16), yielding the following result for the spinodal

$$\tilde{J}_s^{\mathrm{BG}}(\varphi) = \frac{1}{4}\ln\left(\frac{(1+(\bar{z}-2)\varphi)(\bar{z}-1-(\bar{z}-2)\varphi)}{(\bar{z}-2)^2\varphi(1-\varphi)}\right). \tag{2.25}$$

Similarly to the binodal, we find that the spinodal also diverges for $\bar{z}=2$. In Fig. 2.3c the spinodal is shown with the dash-dotted line for a lattice with $\bar{z}=4$.

2.8.3 *Critical Point*

The critical point $(\varphi_{\mathrm{crit}}, \tilde{J}_{\mathrm{crit}})$ denotes the **onset of phase separation**, and is the point where the binodal and spinodal meet. Formally, the critical fraction is determined from $\tilde{\mathrm{f}}'''(\varphi_{\mathrm{crit}})=0$. However, based on the $\mathbb{Z}_2$ symmetry of the Ising model, we know that the critical fraction should be at $\varphi_{\mathrm{crit}} = 1/2$. Plugging this result into Eq. (2.20) or (2.25) finally gives the BG critical coupling

$$\tilde{J}_{\mathrm{crit}}^{\mathrm{BG}} = \frac{1}{2}\ln\left(\frac{\bar{z}}{\bar{z}-2}\right). \tag{2.26}$$

Again, we find that for $\bar{z}=2$ the critical coupling diverges. In Fig. 2.3c the critical point for a lattice with $\bar{z}=4$ is indicated with the black dot.

2.9 Free Energy Minima

When $\tilde{h}=0$ the binodal given by Eq. (2.20) coincides with the minima of $\tilde{\mathrm{f}}_{\mathrm{BG}}(\varphi)$. For $\tilde{h}\neq 0$ this is no longer true, since the free energy minima are now shifted while the binodal remains at the same location as shown in Fig. 2.4. Here we calculate the

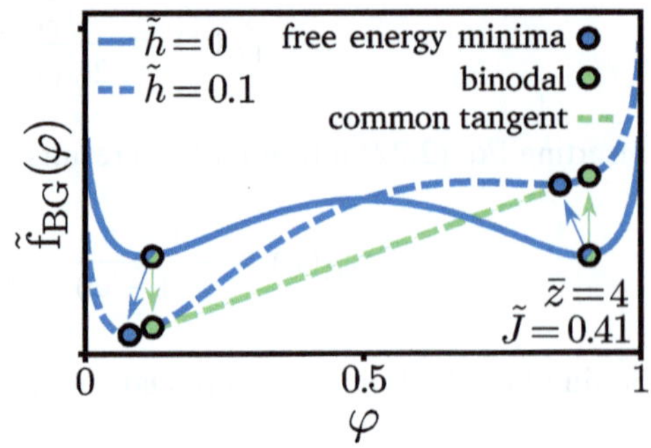

Fig. 2.4 The difference between the binodal (green dots) and free energy minima (blue dots) for $\tilde{h} \neq 0$

free energy minima in the presence of an external field. Based on the analysis in Sect. 2.8.1 we can derive an equation for $\tilde{f}'_{BG}(\varphi) = 0$ including the external field, which reads

$$\chi_\varphi - e^{2\tilde{J}} \left(e^{-2\tilde{h}/\bar{z}} \chi_\varphi^{1-1/\bar{z}} - e^{2\tilde{h}/\bar{z}} \chi_\varphi^{1/\bar{z}} \right) - 1 = 0, \tag{2.27}$$

where $\chi_\varphi \equiv \varphi/(1-\varphi)$. For $\bar{z} = \{2, 3, 4\}$ we can solve Eq. (2.27) analytically. In the next subsections we focus specifically on $\bar{z} = \{2, 4\}$.

2.9.1 One-Dimensional Line with $\bar{z} = 2$

Equation (2.27) can readily be solved for $\bar{z} = 2$. Using the backwards transformation $\varphi = \chi_\varphi/(1+\chi_\varphi)$, we obtain the following result for the free energy minimum, denoted by $\varphi_{\min}^{BG}$

$$\varphi_{\min}^{BG} = \frac{1}{2} \left(1 - \frac{\sinh(\tilde{h})}{\sqrt{e^{-4\tilde{J}} + \sinh(\tilde{h})^2}} \right). \tag{2.28}$$

This result is well-known, and agrees with the exact result obtained by Ising in [9] (see Eq. (2.8) therein), which confirms that the BG approximation is exact on the Bethe lattice.[5]

2.9.2 Square and Bethe Lattice with $\bar{z} = 4$

Upon introducing the auxiliary variable $\xi \equiv \chi_\varphi^{1/4}$ Eq. (2.27) becomes a quartic equation

$$\xi^4 - e^{2\tilde{J}} \left(e^{-\tilde{h}/2} \xi^3 - e^{\tilde{h}/2} \xi \right) - 1 = 0. \tag{2.29}$$

[5] The Bethe lattice for $\bar{z} = 2$ is the one-dimensional line.

The roots of a quartic equation are known, and result in the following solution for Eq. (2.29)

$$\xi_{1,2} = \frac{\mathcal{Y}_+}{4} - S \pm \frac{1}{2}\sqrt{\frac{3}{4}\mathcal{Y}_+^2 + \frac{\mathcal{Y}_- - \frac{1}{8}\mathcal{Y}_+^3}{S} - 4S^2}, \tag{2.30a}$$

$$\xi_{3,4} = \frac{\mathcal{Y}_+}{4} + S \pm \frac{1}{2}\sqrt{\frac{3}{4}\mathcal{Y}_+^2 - \frac{\mathcal{Y}_- - \frac{1}{8}\mathcal{Y}_+^3}{S} - 4S^2}, \tag{2.30b}$$

where we used the following auxiliary variables

$$S = \frac{1}{2}\sqrt{\frac{1}{4}\mathcal{Y}_+^2 + \frac{1}{3}\left(W + \frac{\Delta_0}{W}\right)}, \tag{2.31}$$

$$W = \left(\frac{\Delta_1 + \sqrt{\Delta_1^2 - 4\Delta_0^3}}{2}\right)^{1/3}, \tag{2.32}$$

$$\Delta_0 = 3e^{4\tilde{J}} - 12, \quad \Delta_1 = 54e^{4\tilde{J}}\Delta_{\tilde{h}}, \quad \Delta_{\tilde{h}} = \sinh(\tilde{h}), \quad \mathcal{Y}_\pm = e^{2\tilde{J}\mp\tilde{h}/2}. \tag{2.33}$$

Not all four solutions given by Eqs. (2.30a–2.30b) correspond to the free energy minimum, and therefore we have to carefully analyze them, starting with the discriminant given by

$$\Delta = \frac{1}{27}\left(4\Delta_0^3 - \Delta_1^2\right) = 4(e^{4\tilde{J}} - 4)^3 - 108e^{8\tilde{J}}\Delta_{\tilde{h}}^2. \tag{2.34}$$

For $\Delta < 0$ there are two distinct real roots and two complex conjugate roots, whereas for $\Delta > 0$ there are either four real roots or four imaginary roots, with the former scenario applying here. The discriminant changes sign at the coupling value

$$\tilde{J}_{\Delta=0} = \frac{1}{4}\ln\left(3\Delta_{\tilde{h}}^2(9\Delta_{\tilde{h}}^2 + 8)\Phi^{-1/3} + 3\Phi^{1/3} + 9\Delta_{\tilde{h}}^2 + 4\right), \tag{2.35}$$

with

$$\Phi = 27\Delta_{\tilde{h}}^6 + 36\Delta_{\tilde{h}}^4 + 8\Delta_{\tilde{h}}^2[1 + (1 + \Delta_{\tilde{h}}^2)^{1/2}]. \tag{2.36}$$

Increasing the coupling strength above $\tilde{J}_{\Delta=0}$ gives rise to a local maximum in the free energy landscape, its position being $\varphi_{\max}^{\mathrm{BG}} = \xi_4^4/(1 + \xi_4^4)$ as shown with the purple lines in Fig. 2.5. Furthermore, we find that for $\tilde{h} \neq 0$ the free energy minimum is given by

$$\varphi_{\min}^{\mathrm{BG}}|_{\tilde{h}\neq 0} = \begin{cases} \frac{\xi_1^4}{1+\xi_1^4}, & 0 \leq \tilde{J} \leq \ln\min\{\sqrt{2}, s_0\} \\ \frac{\xi_3^4}{1+\xi_3^4}, & \tilde{J} \geq \ln\min\{\sqrt{2}, s_0\}, \end{cases} . \tag{2.37}$$

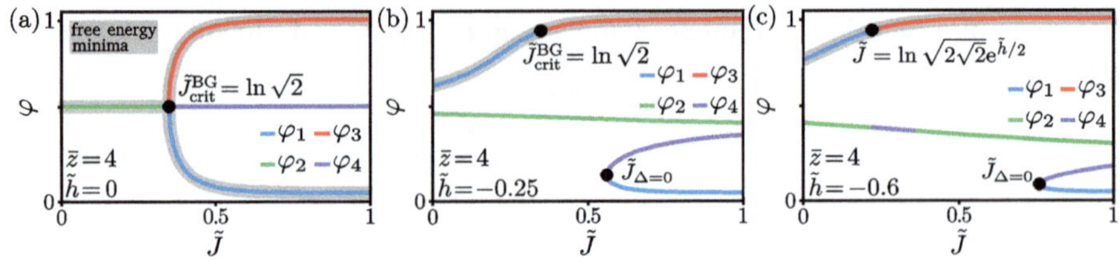

Fig. 2.5 **Stationary points of the BG free energy density**, given by Eqs. (2.30a) and (2.30b), as a function of the coupling strength $\tilde{J}$. From left to right we consider increasing negative values of the external field $\tilde{h}$. The gray shaded lines correspond to the free energy minima given by Eq. (2.38) for $\tilde{h} = 0$ (**a**), and Eq. (2.37) for $\tilde{h} \neq 0$ (**b, c**). The coupling value $\tilde{J}_{\Delta=0}$, given by Eq. (2.35), indicates the coupling strength where all four solutions become real, and denotes the first appearance of a local maximum given by φ_4

where $\ln s_0 \equiv \ln\sqrt{2\sqrt{2}e^{\tilde{h}/2}}$ is the coupling strength that solves for the root of S in Eq. (2.31). In Fig. 2.5b, c the free energy minimum is indicated with the gray shaded blue and red line.

For $\tilde{h} = 0$ the four solutions (2.30a and 2.30b) simplify substantially, and result in the following outcome for the minima (see also Fig. 2.5a)

$$\varphi^{\rm BG}_{\rm min}|_{\tilde{h}=0} = \begin{cases} \frac{1}{2}, & 0 \leq \tilde{J} \leq \ln\sqrt{2} \\ \frac{1}{2}\left[1 \pm \frac{e^{2\tilde{J}}\sqrt{e^{4\tilde{J}}-4}}{e^{4\tilde{J}}-2}\right], & \tilde{J} \geq \ln\sqrt{2} \end{cases} . \tag{2.38}$$

Equation (2.38) disagrees with Onsager's exact result for the square lattice given by Eq. (1.5), but agrees with the spontaneous magnetization on the Bethe lattice with $\bar{z} = 4$.

2.10 Mean Field Approximation

Here we list the results obtained with the MF approximation for sake of completeness. Notably, the BG results converge to the MF results for $\tilde{J} \to 0$ or $\bar{z} \to \infty$, where they are both exact.

2.10.1 Fixed-Magnetization Partition Function

The MF fixed-magnetization partition function is easily obtained from the BG expression by setting $\zeta^{\dagger}_{\rm BG}(\varphi) = \zeta_{\rm BG}(\varphi)$ in Eq. (2.13), which gives

$$\mathcal{Z}^{\rm MF}_{k=N\varphi} = \binom{N}{N\varphi} e^{-2\bar{z}N\tilde{J}(\varphi(1-\varphi)-1/4)}. \tag{2.39}$$

2.10.2 Free Energy Density

Inserting Eq. (2.39) into Eq. (2.2) we obtain the MF free energy density (where $\Xi(\varphi) \equiv \varphi \ln \varphi$)

$$\tilde{\mathrm{f}}_{\mathrm{MF}}(\varphi) = 2\bar{z}\tilde{J}[\varphi(1-\varphi) - 1/4] - \tilde{h}(1-2\varphi) + \Xi(\varphi) + \Xi(1-\varphi). \tag{2.40}$$

2.10.3 Phase Diagram

The binodal, spinodal, and critical point obtained with the MF approximation read

$$\tilde{J}_b^{\mathrm{MF}}(\varphi) = \frac{1}{2\bar{z}(2\varphi-1)} \ln\left(\frac{\varphi}{1-\varphi}\right), \ \tilde{J}_s^{\mathrm{MF}}(\varphi) = \frac{1}{4\bar{z}} \frac{1}{\varphi(1-\varphi)}, \ \tilde{J}_{\mathrm{crit}}^{\mathrm{MF}} = \frac{1}{\bar{z}}. \tag{2.41}$$

These results are well known and are extensively reported in the literature [2–4, 10, 11].

2.10.4 Free Energy Minima

The free energy minima of the MF free energy density can be written as $\varphi_{\mathrm{min}}^{\mathrm{MF}} = (1+s)/2$, where s are the non-zero solutions of the transcendental equation

$$s = \tanh(\bar{z}\tilde{J}s - \tilde{h}). \tag{2.42}$$

For $\tilde{h} = 0$, and below the critical coupling $\tilde{J} \leq 1/\bar{z}$, the only solution to Eq. (2.42) is given by $s = 0$. For $\tilde{J} > 1/\bar{z}$ there exists two nonzero solutions resulting in $\varphi_{\mathrm{min}}^{\mathrm{MF}} \neq 1/2$.

2.11 Qualitative Differences Between MF and BG Approximation

2.11.1 Free Energy Barrier and Curvature

It has already been stated in Chap. 1 that the MF approximation erroneously predicts spontaneous magnetization in one dimension. Yet, there are more subtle differences between the BG and MF approximation, which become apparent in the strong coupling limit. In Fig. 2.6a we plot the MF (red) and BG (blue) free energy density for $\bar{z} = 4$ with $(\tilde{J}, \tilde{h}) = (0.5, 0)$. Two characteristic differences between the two free

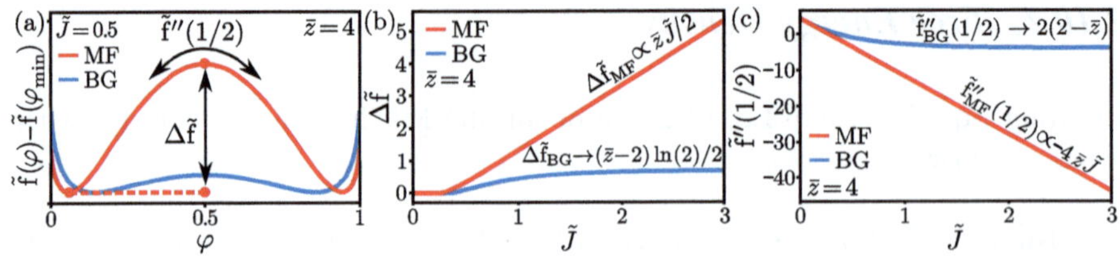

Fig. 2.6 Qualitative differences between the MF and BG free energy density. In each plot we consider a lattice with $\bar{z} = 4$ and vanishing external field $\tilde{h} = 0$. Red lines correspond to the MF approximation and blue lines to the BG approximation. **a** Free energy density as a function of the concentration of down spins φ at fixed $\tilde{J} = 0.5$. $\Delta\tilde{f} \equiv \tilde{f}(1/2) - \tilde{f}(\varphi_{\min})$ denotes the free energy barrier between the two minima, and $\tilde{f}''(1/2)$ is the curvature of the free energy barrier. **b** $\Delta\tilde{f}$ as a function of $\tilde{J}$. The MF free energy barrier diverges to $+\infty$, whereas the BG barrier converges to $\lim_{\tilde{J}\to\infty} \Delta\tilde{f}_{BG} = (\bar{z} - 2)\ln(2)/2$. **c** $\tilde{f}''(1/2)$ as a function of $\tilde{J}$. The MF curvate diverges to $-\infty$, whereas the BG curvature converges to $\lim_{\tilde{J}\to\infty} \tilde{f}''_{BG}(1/2) = 2(2 - \bar{z})$

energy densities are the height of the **free energy barrier** $\Delta\tilde{f} \equiv \tilde{f}(1/2) - \tilde{f}(\varphi_{\min})$, and the **curvature of the barrier** $\tilde{f}''(1/2)$, indicated with the black arrows. In Fig. 2.6b, c both $\Delta\tilde{f}$ and $\tilde{f}''(1/2)$ are plotted w.r.t. $\tilde{J}$, and we find that the MF results diverge in stark contrast to the BG results. The barrier and curvature determine nucleation times, stable wavelengths in spinodal decomposition, stability of uniform mixtures, and the shape of concentration profiles [5, 12, 13]. Therefore, based on Fig. 2.6b, c, we find that these physical observables should have clear distinct qualitative behavior between the BG and MF approximation, which will be further discussed in Chaps. 4–5. To obtain the free energy barrier and curvature for $\tilde{J} \to \infty$ we use $\lim_{\tilde{J}\to\infty} \varphi_{\min} = 0 \vee 1$ for $\bar{z} \geq 3$[6] and $\tilde{h} = 0$, which gives the following results using Eqs. (2.14) and (2.40)

$$\begin{aligned}
\lim_{\tilde{J}\to\infty} \Delta\tilde{f}_{BG} &= (\bar{z} - 2)\ln(2)/2, & \lim_{\tilde{J}\to\infty} \Delta\tilde{f}_{MF} &= \bar{z}\tilde{J}/2, \\
\lim_{\tilde{J}\to\infty} \tilde{f}''_{BG}(1/2) &= 2(2 - \bar{z}), & \lim_{\tilde{J}\to\infty} \tilde{f}''_{MF}(1/2) &= -4\bar{z}\tilde{J}. \qquad (2.43)
\end{aligned}$$

Both results are displayed in Fig. 2.6b, c above their respective lines.

2.11.2 Error Analysis

To systematically asses the accuracy of the MF and BG approximation, we want to compare them with exact results. Let $\tilde{f}_N(\varphi)$ be the free energy density for a finite system composed of N spins. Using Eqs. (2.2) and (1.6), we can calculate $\tilde{f}_N(\varphi)$ numerically exact by brute-force summation of the fixed-magnetization partition function. In Fig. 2.7 we compare the accuracy of the BG (a) and MF (b) approximations by means of the relative error $\epsilon_N^{BG,MF}(\varphi) \equiv 1 - \tilde{f}_N^{BG/MF}(\varphi)/\tilde{f}_N(\varphi)$ for a square

[6] For the MF approximation this statement also holds for $\bar{z} = 2$.

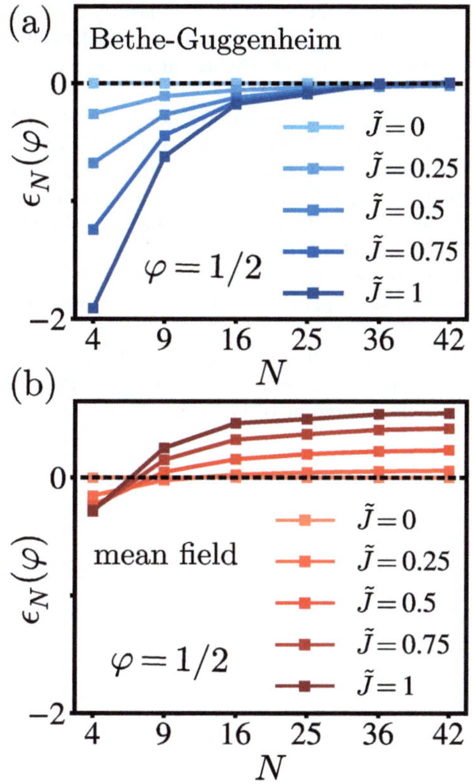

Fig. 2.7 Relative error of the free energy density for $\tilde{h}=0$ obtained with the (**a**) BG and (**b**) MF approximation as a function of the system size. Up to $N=36$ each point is calculated for a square lattice of size $\sqrt{N}\times\sqrt{N}$ with free boundary conditions and $\varphi=1/2$. For $N=42$ a rectangular lattice of size 6×7 is considered

lattice ($\bar{z}=4$) with $\varphi=1/2$ closed bonds and varying size N. We find that the BG approximation approaches the numerically exact free energy density with increasing N, regardless of the coupling strength $\tilde{J}$. Conversely, the MF approximation diverges with both, increasing system size N and coupling strength $\tilde{J}$.

References

1. Longuet-Higgins HC, Fisher ME (1978) Lars Onsager, 27 November–5 October 1976. Biogr Mem Fellows R Soc 24:443–471. [Online]. Available https://royalsocietypublishing.org/doi/abs/10.1098/rsbm.1978.0014
2. Fowler RH (1939) Statistical thermodynamics. CUP Archive
3. Guggenheim EA (1952) Mixtures: the theory of the equilibrium properties of some simple classes of mixtures solutions and alloys. Clarendon Press
4. de With G (2013) Liquid-state physical chemistry: fundamentals, modeling, and applications. Wiley-VCH Verlag
5. Blom K, Godec A (2021) Criticality in cell adhesion. Phys Rev X 11(3):031067. [Online]. Available https://link.aps.org/doi/10.1103/PhysRevX.11.031067

6. Saito Y, Kubo R (1976) Nonlinearity in cooperative systems-dynamical Bethe-Ising model. J Stat Phys 15:233–253. [Online]. Available https://doi.org/10.1007/BF01012879
7. McQuarrie DA (1976) Statistical mechanics. In: Harper's chemistry series. HarperCollins Publishing Inc., New York
8. Hill TL (1986) An introduction to statistical thermodynamics. Courier Corporation
9. Ising E (1925) Beitrag zur Theorie des Ferromagnetismus. Zeitschrift für Physik 31(1):253–258. [Online]. Available http://link.springer.com/10.1007/BF02980577
10. Doi M (2013) Soft matter physics. Oxford University Press
11. Weber CA, Zwicker D, Jülicher F et al (2019) Physics of active emulsions. Rep Prog Phys 82(6):064601. [Online]. Available https://doi.org/10.1088/1361-6633/ab052b
12. Cahn JW, Hilliard JE (1958) Free energy of a nonuniform system. I. Interfacial free energy. J Chem Phys 28(2):258–267. [Online]. Available https://aip.scitation.org/doi/10.1063/1.1744102
13. Blom K, Ziethen N, Godec A (2022) Thermodynamically consistent phase-field theory including nearest-neighbor pair correlations. Phys Rev Res 5(1):013135. [Online]. Available https://link.aps.org/doi/10.1103/PhysRevResearch.5.013135

Chapter 3
Bethe-Guggenheim Approximation for Non-uniform Systems

Well, I come down in the morning and I take up a pencil and I try to THINK.

Hans Bethe in Hans Bethe: Prophet of Energy

Building upon the previous chapter, we here apply the BG approximation to non-uniform lattice systems. We consider an Ising model with anisotropic coupling (J_x, J_y) in the horizontal and vertical direction and calculate the partition function for a fixed magnetization per spin block. Upon taking the thermodynamic limit, we obtain a **Cahn-Hilliard free energy functional** with a **concentration-dependent gradient energy coefficient**. Subsequently, we determine the equilibrium concentration profile and stable wavelength perturbations within the spinodal region. Here, we encounter a remarkable new feature of the BG concentration profile named **interface broadening**. This phenomenon is not present in the MF approximation, and will be further investigated in Chap. 4. In the final section we provide a comparison between the BG and MF results, and show how their concentration profiles differ qualitatively in the strong coupling limit. Part of this chapter has been published in *Physical Review Research* under the terms of the Creative Commons Attribution 4.0 International license [1].

3.1 Initial Setup and Definitions

3.1.1 Lattice Specification and the Thermodynamic Limit

In this chapter the horizontal and vertical direction are indicated with x and y, respectively.

K. Blom, *Pair-Correlation Effects in Many-Body Systems*, Springer Theses,
https://doi.org/10.1007/978-3-031-29612-3_3

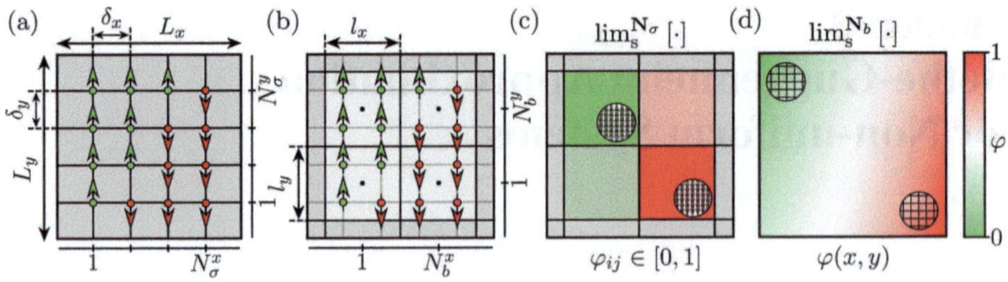

Fig. 3.1 Lattice specification and the thermodynamic limits. a, b Lattice configuration of the spins (**a**), and spin blocks (**b**). $(L_{x,y}, l_{x,y}, \delta_{x,y})$ are the lattice length, spin block length, and lattice spacing, respectively. The number of spins and spin blocks are denoted with $(N_\sigma^{x,y}, N_b^{x,y})$. Here we consider an example with sixteen spins and four spin blocks. **c, d** Thermodynamic limit of the spins (**c**) and spin blocks (**d**) defined in Eq. (3.1). The circles display a magnification of individual spins (**c**) and spin blocks (**d**)

3.1.1.1 Spins

We consider $N = N_\sigma^x \times N_\sigma^y$ spins $\sigma_{ij} = \pm 1$ with $(i, j) \in (\{1, ..., N_\sigma^x\}, \{1, ..., N_\sigma^y\})$ arranged on a lattice with sides (L_x, L_y). In Fig. 3.1a we provide an example for a square lattice with sixteen spins. The lattice spacings between spins are $(\delta_x, \delta_y) = (L_x/N_\sigma^x, L_y/N_\sigma^y)$. The average lattice coordination number is denoted with $\bar{z} = \bar{z}_x + \bar{z}_y$, and $\bar{\mathbf{z}} = \text{diag}(\bar{z}_x, \bar{z}_y)$ is a diagonal matrix containing the average lattice coordination numbers in the horizontal and vertical direction, respectively. The square lattice in Fig. 3.1 has $(\bar{z}_x, \bar{z}_y) = (2, 2)$ and $\bar{z} = 4$.

3.1.1.2 Spin Blocks

Similar to Kadanoff's block spin method [2] we place the spins into $N_b = N_b^x \times N_b^y$ blocks as shown in Fig. 3.1b. Let $\mathbf{b}_{ij}$ with $(i, j) \in (\{1, ..., N_b^x\}, \{1, ..., N_b^y\})$ denote such a block containing $\dim \mathbf{b}_{ij} = N_\sigma^x/N_b^x \times N_\sigma^y/N_b^y$ spins. Consequently, the horizontal and vertical length of each block is given by $(l_x, l_y) = (L_x/N_b^x, L_y/N_b^y)$. The blocks have the same average lattice coordination number as the spins. In Fig. 3.1b each block consists of four spins and is aligned with 2 blocks in the horizontal and vertical direction, respectively.

3.1.1.3 Thermodynamic Limit

As in Chap. 2, we will construct a field theory for the fraction of down spins φ. To that aim we introduce the following two scaling limits where we take the number of spins/blocks to infinity while simultaneously keeping the block/lattice length fixed, i.e.

$$\begin{aligned}\lim_{\mathrm{s}}^{\mathbf{N}_\sigma}[\cdot] &\equiv \lim_{\substack{N_\sigma^x, N_\sigma^y \to \infty \\ l_x, l_y = (N_\sigma^x/N_b^x)\delta_x, (N_\sigma^y/N_b^y)\delta_y = \text{const.}}} [\cdot]\,, \\ \lim_{\mathrm{s}}^{\mathbf{N}_b}[\cdot] &\equiv \lim_{\substack{N_b^x, N_b^y \to \infty \\ L_x, L_y = N_b^x l_x, N_b^y l_y = \text{const.}}} [\cdot]\,,\end{aligned} \tag{3.1}$$

where $\mathbf{N}_\sigma$ and $\mathbf{N}_b$ denote the thermodynamic limit of the spins and blocks, respectively. In Fig. 3.1c–d we give a schematic representation of both limits.

3.1.2 Coarse-Grained Lattice Observables

Recall that dim $\mathbf{b}_{ij}$ denotes the number of spins inside each spin block. Let us define the fraction of down spins in block $\mathbf{b}_{ij}$ as

$$\varphi_{ij}(\mathbf{b}_{ij}) \equiv (\dim \mathbf{b}_{ij})^{-1} \sum_{(m,n)\in\mathbf{b}_{ij}} (1 - \sigma_{mn})/2. \tag{3.2}$$

For a finite number of spins within each block φ_{ij} is a rational number. Applying the first scaling limit in Eq. (3.1) takes the number of spins within each block to infinity, rendering $\lim_{\mathrm{s}}^{\mathbf{N}_\sigma}\left[\varphi_{ij}\right] \in [0, 1]$ a continuous variable. The second limit takes the number of blocks to infinity while simultaneously decreasing their distance, resulting in a continuous differentiable field $\lim_{\mathrm{s}}^{\mathbf{N}_b}\left[\lim_{\mathrm{s}}^{\mathbf{N}_\sigma}\left[\varphi_{ij}\right]\right] = \varphi(x, y)$ as depicted in Fig. 3.1d.

Additionally to Eq. (3.2), we need to define the fraction of **intrablock defects** inside $\mathbf{b}_{ij}$ in the horizontal and vertical direction, which are respectively given by

$$\begin{aligned}\zeta_{ij}^x(\mathbf{b}_{ij}) &\equiv (\bar{z}_x \dim \mathbf{b}_{ij})^{-1} \sum_{(m,n)\in\mathbf{b}_{ij}} \sum_{\substack{(k,n)\in\mathbf{b}_{ij} \\ \langle mk\rangle}} |\sigma_{mn} - \sigma_{kn}|/2, \\ \zeta_{ij}^y(\{\mathbf{b}_{ij}\}) &\equiv (\bar{z}_y \dim \mathbf{b}_{ij})^{-1} \sum_{(m,n)\in\mathbf{b}_{ij}} \sum_{\substack{(m,k)\in\mathbf{b}_{ij} \\ \langle nk\rangle}} |\sigma_{mn} - \sigma_{mk}|/2,\end{aligned} \tag{3.3}$$

where $\langle nk\rangle$ denotes nearest neighbors. Finally, we define the fraction of **interblock defects** between neighbouring blocks in the horizontal and vertical direction, respectively given by

$$\begin{aligned}\xi_{ij}^{x\pm}(\{\mathbf{b}_{ij}, \mathbf{b}_{i\pm1j}\}) &\equiv (\bar{z}_x \dim \mathbf{b}_{ij})^{-1} \sum_{(m,n)\in\mathbf{b}_{ij}} \sum_{\substack{(k,n)\in\mathbf{b}_{i\pm1j} \\ \langle mk\rangle}} |\sigma_{mn} - \sigma_{kn}|/2, \\ \xi_{ij}^{y\pm}(\{\mathbf{b}_{ij}, \mathbf{b}_{ij\pm1}\}) &\equiv (\bar{z}_y \dim \mathbf{b}_{ij})^{-1} \sum_{(m,n)\in\mathbf{b}_{ij}} \sum_{\substack{(m,k)\in\mathbf{b}_{ij\pm1} \\ \langle nk\rangle}} |\sigma_{mn} - \sigma_{mk}|/2,\end{aligned} \tag{3.4}$$

where one needs to account for the boundary conditions upon summing over boundary blocks. As an example, we calculate the fraction of intra- and interblock defects for the lower left block in Fig. 3.1b, which gives for periodic boundary conditions: $\zeta_{11}^{x,y} = 1/8$, $\xi_{11}^{x+} = 1/8$, $\xi_{11}^{x-} = 2/8$, $\xi_{11}^{y+} = 0$, and $\xi_{11}^{y-} = 1/8$.

3.2 Rewriting the Ising Hamiltonian over Spin Blocks

We now write the nearest neighbour Ising Hamiltonian in terms of the coarse-grained intensive lattice variables introduced in Eqs. (3.2)–(3.4). Our starting point is the representation of the Ising Hamiltonian in terms of defects given by Eq. (2.5) in Chap. 2. Let the spin interaction strength in the horizontal and vertical direction be given by $(\tilde{J}_x, \tilde{J}_y) \equiv \beta(J_x, J_y)$, where $\beta \equiv 1/k_{\mathrm{B}}T$ is the inverse thermal energy. We decompose the Hamiltonian given by Eq. (2.5) into a sum over inter- and intrablock contributions (recall that $\tilde{\mathcal{H}} \equiv \beta\mathcal{H}$)

$$\tilde{\mathcal{H}}(\boldsymbol{b}) = \dim \mathbf{b}_{ij} \left(\sum_{i=1}^{N_b^x} \sum_{j=1}^{N_b^y} [\tilde{\mathcal{H}}_{\mathrm{inter}}(\{\mathbf{b}_{i\pm 1j}, \mathbf{b}_{ij}, \mathbf{b}_{ij\pm 1}\})/2 + \tilde{\mathcal{H}}_{\mathrm{intra}}(\{\mathbf{b}_{ij}\}) - \mathcal{C}] \right). \tag{3.5}$$

The respective contributions inside the sum are given by

$$\begin{aligned} \tilde{\mathcal{H}}_{\mathrm{inter}}(\{\mathbf{b}_{i\pm 1j}, \mathbf{b}_{ij}, \mathbf{b}_{ij\pm 1}\}) &= \bar{z}_x \tilde{J}_x [\xi_{ij}^{x+}(\{\mathbf{b}_{ij}, \mathbf{b}_{i+1j}\}) + \xi_{ij}^{x-}(\{\mathbf{b}_{ij}, \mathbf{b}_{i-1j}\})] + (x \leftrightarrow y), \\ \tilde{\mathcal{H}}_{\mathrm{intra}}(\{\mathbf{b}_{ij}\}) &= \bar{z}_x \tilde{J}_x \zeta_{ij}^{x}(\{\mathbf{b}_{ij}\}) + (x \leftrightarrow y), \end{aligned} \tag{3.6}$$

and the constant reads $\mathcal{C} \equiv (\bar{z}_x \tilde{J}_x + \bar{z}_y \tilde{J}_y)/2$. The term $(x \leftrightarrow y)$ in Eq. (3.6) denotes a repetition of the preceding term with x and y interchanged. Note that the factor $1/2$ for $\tilde{\mathcal{H}}_{\mathrm{inter}}$ in Eq. (3.5) accounts for the double counting over interblock contributions.

3.3 Rewriting the Partition Function over Spin Blocks

Inserting Eq. (3.5) in combination with Eq. (3.6) into the partition function and restoring the external field $\tilde{h}$ leads to the following *exact* expression

$$\mathcal{Z} = \sum_{\{\mathbf{b}_{ij}\}} \prod_{i=1}^{N_b^x} \prod_{j=1}^{N_b^y} \mathrm{e}^{-\dim \mathbf{b}_{ij} [\tilde{\mathcal{H}}_{\mathrm{inter}}(\{\mathbf{b}_{ij}, \mathbf{b}_{i\pm 1j}, \mathbf{b}_{ij\pm 1}\})/2 + \tilde{\mathcal{H}}_{\mathrm{intra}}(\{\mathbf{b}_{ij}\}) - \tilde{h}(1 - 2\varphi_{ij}(\{\mathbf{b}_{ij}\})) - \mathcal{C}]}. \tag{3.7}$$

Note that we are currently not constraining the magnetization. The evaluation of the configurational sum over all possible spin block configurations $\{\mathbf{b}_{ij}\}$ constitutes a difficult—if not impossible—task. However, the Hamiltonian inside the sum solely depends on the seven lattice variables $(\varphi_{ij}, \zeta_{ij}^{x,y}, \xi_{ij}^{x,y\pm})$ $(i, j) \in (\{1, ..., N_b^x\}, \{1, ..., N_b^y\})$. Therefore, we can interchange the configurational sum by a sum over all possible values of these seven lattice variables and introduce a **degeneracy of states** $\Psi(\varphi_{ij}, \zeta_{ij}^{x,y}, \xi_{ij}^{x,y\pm})$ which accounts for the multiplicity of configurations. Hence

$$\mathcal{Z} = \sum_{\varphi_{ij}} \sum_{\zeta_{ij}^{x,y}} \sum_{\xi_{ij}^{x,y\pm}} \prod_{i=1}^{N_b^x} \prod_{j=1}^{N_b^y} \Psi(\varphi_{ij}, \zeta_{ij}^{x,y}, \xi_{ij}^{x,y\pm}) \mathrm{e}^{-\dim \mathbf{b}_{ij} [\bar{z}_x \tilde{J}_x [\zeta_{ij}^x + (\xi_{ij}^{x+} + \xi_{ij}^{x-})/2] + (x \leftrightarrow y) - \tilde{h}(1-2\varphi_{ij}) - \mathcal{C}]}. \tag{3.8}$$

Equation (3.8) remains an exact expression as long as the degeneracy of states $\Psi(\varphi_{ij}, \zeta_{ij}^{x,y}, \xi_{ij}^{x,y\pm})$ is evaluated correctly. For $(\tilde{J}, \tilde{h}) = (0, 0)$ we know that the degeneracy of states should obey the following relation

$$\sum_{\zeta_{ij}^{x,y}} \sum_{\xi_{ij}^{x,y\pm}} \Psi(\varphi_{ij}, \zeta_{ij}^{x,y}, \xi_{ij}^{x,y\pm}) \stackrel{!}{=} \binom{\dim \mathbf{b}_{ij}}{\varphi_{ij} \dim \mathbf{b}_{ij}}, \tag{3.9}$$

since this is the amount of possible configurations to place $\varphi_{ij} \dim \mathbf{b}_{ij}$ down spins in a block that contains $\dim \mathbf{b}_{ij}$ spins in total. We will use Eq. (3.9) as a normalization condition to consistently approximate the degeneracy of states.

3.4 Pair Approximation Ansatz

Similar to Chap. 2, we introduce a pair-approximation Ansatz for the degeneracy of states. The difference, however, is that now we account for the difference between intra- and interblock contributions. This results in the following expression for the degeneracy of states for a non-uniform system

$$\begin{aligned} &\Psi_{\mathrm{BG}}(\varphi_{i\pm1j}, \varphi_{ij}, \varphi_{ij\pm1}, \zeta_{ij}^{x,y}, \xi_{ij}^{x,y\pm}) \\ &= \mathcal{N}_{\mathrm{BG}}(\varphi_{i\pm1j}, \varphi_{ij}, \varphi_{ij\pm1}) \hat{\Psi}_{\mathrm{BG}}(\varphi_{i\pm1j}, \varphi_{ij}, \varphi_{ij\pm1}, \zeta_{ij}^{x,y}, \xi_{ij}^{x,y\pm}), \end{aligned} \tag{3.10}$$

where $\mathcal{N}_{\mathrm{BG}}(\varphi_{i\pm1j}, \varphi_{ij}, \varphi_{ij\pm1})$ is a normalization constant left to be determined. The unnormalized degeneracy of states reads

$$\begin{aligned} &\hat{\Psi}_{\mathrm{BG}}(\varphi_{i\pm1j}, \varphi_{ij}, \varphi_{ij\pm1}, \zeta_{ij}^{x,y}, \xi_{ij}^{x,y\pm}) \\ &= \hat{\Psi}_{\mathrm{BG,intra}}(\varphi_{ij}, \zeta_{ij}^{x,y}) \hat{\Psi}_{\mathrm{BG,inter}}(\varphi_{i\pm1j}, \varphi_{ij}, \varphi_{ij\pm1}, \xi_{ij}^{x,y\pm}), \end{aligned} \tag{3.11}$$

which we have further split into intra- and interblock contributions, respectively, given by

$$\hat{\Psi}_{\mathrm{BG,intra}}(\varphi_{ij}, \zeta_{ij}^{x,y}) = \psi_x(\varphi_{ij}, \varphi_{ij}, \zeta_{ij}^{x})\psi_y(\varphi_{ij}, \varphi_{ij}, \zeta_{ij}^{y}), \tag{3.12}$$

$$\hat{\Psi}_{\mathrm{BG,inter}}(\varphi_{i\pm1j}, \varphi_{ij}, \varphi_{ij\pm1}, \xi_{ij}^{x,y\pm}) = \prod_{\pm} \psi_x^{\frac{1}{2}}(\varphi_{ij}, \varphi_{i\pm1j}, \xi_{ij}^{x\pm})\psi_y^{\frac{1}{2}}(\varphi_{ij}, \varphi_{ij\pm1}, \xi_{ij}^{y\pm}). \tag{3.13}$$

The functions $\psi_{x,y}(a, b, c)$, which enter Eqs. (3.12) and (3.13), are given by

$$\psi_{x,y}(a, b, c) \equiv [\tilde{\Gamma}_{x,y}(1 - a - c)\tilde{\Gamma}_{x,y}(b - c)\tilde{\Gamma}_{x,y}(a - b + c)\tilde{\Gamma}_{x,y}(c)]^{-1/2}, \tag{3.14}$$

with $\tilde{\Gamma}_{x,y}(w) \equiv \Gamma(\bar{z}_{x,y}(\dim \mathbf{b}_{ij})w/2 + 1)$, and $\Gamma(w)$ being the Gamma function (i.e. generalized factorial). Equation (3.14) can be derived similarly to Eq. (2.9) in Chap. 2 by counting the number of degenerate configurations upon distributing pairs of spins over a lattice. The functions $\psi_{x,y}(\varphi_{ij}, \varphi_{ij}, \zeta_{ij}^{x,y})$ account for distributing pairs of spins inside a single block in the horizontal and vertical direction, respectively. Similarly, $\psi_x(\varphi_{ij}, \varphi_{i\pm1j}, \xi_{ij}^{x\pm})$ and $\psi_y(\varphi_{ij}, \varphi_{ij\pm1}, \xi_{ij}^{y\pm})$ account for distributing pairs of spins between two neighbouring blocks in the horizontal and vertical direction, respectively.

3.5 Evaluating the Normalization Constant

The normalization constant $\mathcal{N}_{\mathrm{BG}}(\varphi_{i\pm1j}, \varphi_{ij}, \varphi_{ij\pm1})$ is determined by equation Eq. (3.9). To evaluate the six sums over the lattice variables $(\zeta_{ij}^{x,y}, \xi_{ij}^{x,y\pm})$, we take the thermodynamic limit of the spins, which renders the lattice variables continuous, and employ the maximum term method. This results in the following arguments

$$\begin{aligned}
\zeta_{\mathrm{BG}}(\varphi_{ij}, \varphi_{ij}) &\equiv \arg\sup_{\zeta_{ij}^{x,y}}\{\lim_{\mathrm{s}}^{\mathbf{N}_\sigma}[(\dim \mathbf{b}_{ij})^{-1}\ln(\psi_{x,y}(\varphi_{ij}, \varphi_{ij}, \zeta_{ij}^{x,y}))]\}\\
&= \varphi_{ij}(1 - \varphi_{ij}),\\
\xi_{\mathrm{BG}}^{x\pm}(\varphi_{i\pm1j}, \varphi_{ij}) &\equiv \arg\sup_{\xi_{ij}^{x\pm}}\{\lim_{\mathrm{s}}^{\mathbf{N}_\sigma}[(\dim \mathbf{b}_{ij})^{-1}\ln(\psi_x(\varphi_{ij}, \varphi_{i\pm1j}, \xi_{ij}^{x\pm}))]\}\\
&= \varphi_{i\pm1j}(1 - \varphi_{ij}),\\
\xi_{\mathrm{BG}}^{y\pm}(\varphi_{ij\pm1}, \varphi_{ij}) &\equiv \arg\sup_{\xi_{ij}^{y\pm}}\{\lim_{\mathrm{s}}^{\mathbf{N}_\sigma}[(\dim \mathbf{b}_{ij})^{-1}\ln(\psi_y(\varphi_{ij}, \varphi_{ij\pm1}, \xi_{ij}^{y\pm}))]\}\\
&= \varphi_{ij\pm1}(1 - \varphi_{ij}).
\end{aligned} \tag{3.15}$$

To acquire Eq. (3.15), we have used Stirling's approximation for the Gamma function $\ln \Gamma(w) = \Xi(w) - w + \mathcal{O}(\ln w)$ for $\mathrm{Re}(w) > 0$ with $\Xi(w) \equiv w \ln(w)$. Plugging Eq. (3.15) into Eq. (3.10), and finally solving Eq. (3.9) for the normalization constant yields

$$\mathcal{N}_{\mathrm{BG}}(\varphi_{i\pm 1j}, \varphi_{ij}, \varphi_{ij\pm 1}) = \binom{\dim \mathbf{b}_{ij}}{\varphi_{ij}\dim \mathbf{b}_{ij}} \hat{\Psi}_{\mathrm{BG}}^{-1}(\varphi_{i\pm 1j}, \varphi_{ij}, \varphi_{ij\pm 1}, \zeta_{\mathrm{BG}}, \xi_{\mathrm{BG}}^{x,y\pm}), \quad (3.16)$$

where we have suppressed the arguments of ζ_{BG} and $\xi_{\mathrm{BG}}^{x,y\pm}$ for convenience. With the normalization constant evaluated explicitly, we can now determine the partition function given by Eq. (3.8).

3.6 Evaluating the Partition Sum

Similarly to the previous section we take the thermodynamic limit of the spins and approximate the six inner sums in Eq. (3.8) with the maximum term method, which yields

$$\begin{aligned} \hat{\zeta}_{\mathrm{BG}}^{x,y}(\varphi_{ij}, \varphi_{ij}) &\equiv \arg\sup_{\zeta_{ij}^{x,y}} \left\{ \lim_{\mathrm{s}}^{\mathbf{N}_\sigma} \left[(\dim \mathbf{b}_{ij})^{-1} \ln \left(\psi_{x,y}(\varphi_{ij}, \varphi_{ij}, \zeta_{ij}^{x,y}) \mathrm{e}^{-\bar{z}_{x,y} \tilde{J}_{x,y} \zeta_{ij}^{x,y} \dim \mathbf{b}_{ij}} \right) \right] \right\} \\ &= \frac{2\zeta_{\mathrm{BG}}}{\Omega_{x,y}(0, \zeta_{\mathrm{BG}})}, \end{aligned} \quad (3.17)$$

$$\begin{aligned} \hat{\xi}_{\mathrm{BG}}^{x\pm}(\varphi_{i\pm 1j}, \varphi_{ij}) &\equiv \arg\sup_{\xi_{ij}^{x\pm}} \left\{ \lim_{\mathrm{s}}^{\mathbf{N}_\sigma} \left[(\dim \mathbf{b}_{ij})^{-1} \ln \left(\psi_x^{\frac{1}{2}}(\varphi_{ij}, \varphi_{i\pm 1j}, \xi_{ij}^{x\pm}) \mathrm{e}^{-\bar{z}_x \tilde{J}_x \xi_{ij}^{x\pm} \dim \mathbf{b}_{ij}/2} \right) \right] \right\} \\ &= \frac{2\xi_{\mathrm{BG}}^{x\pm}}{\Omega_x(\varphi_{ij} - \varphi_{i\pm 1j}, \xi_{\mathrm{BG}}^{x\pm})}, \end{aligned} \quad (3.18)$$

$$\begin{aligned} \hat{\xi}_{\mathrm{BG}}^{y\pm}(\varphi_{ij\pm 1}, \varphi_{ij}) &\equiv \arg\sup_{\xi_{ij}^{y\pm}} \left\{ \lim_{\mathrm{s}}^{\mathbf{N}_\sigma} \left[(\dim \mathbf{b}_{ij})^{-1} \ln \left(\psi_y^{\frac{1}{2}}(\varphi_{ij}, \varphi_{ij\pm 1}, \xi_{ij}^{y\pm}) \mathrm{e}^{-\bar{z}_y \tilde{J}_y \xi_{ij}^{y\pm} \dim \mathbf{b}_{ij}/2} \right) \right] \right\} \\ &= \frac{2\xi_{\mathrm{BG}}^{y\pm}}{\Omega_y(\varphi_{ij} - \varphi_{ij\pm 1}, \xi_{\mathrm{BG}}^{y\pm})}, \end{aligned} \quad (3.19)$$

where we introduced the auxiliary function

$$\Omega_{x,y}(a, b) \equiv 1 + a\gamma_{x,y} + [\delta_{a,0} + \mathrm{sgn}(a)]([1 + a\gamma_{x,y}]^2 + 4b\gamma_{x,y})^{\frac{1}{2}}, \quad (3.20)$$

with $\mathrm{sgn}(x) = \pm 1$ for $\pm x > 0$, $\mathrm{sgn}(0) = 0$, and $\gamma_{x,y} \equiv \exp(4\tilde{J}_{x,y}) - 1$. For $\tilde{J}_{x,y} = 0$ we have $\hat{\zeta}_{\mathrm{BG}}^{x,y} = \zeta_{\mathrm{BG}}$ and $\hat{\xi}_{\mathrm{BG}}^{x,y\pm} = \xi_{\mathrm{BG}}^{x,y\pm}$, as expected. Note that Eq. (3.17) equals Eq. (2.12) in Chap. 2. With the six inner sums evaluated the BG partition function becomes

$$
\begin{aligned}
\mathcal{Z}_{\mathrm{BG}} &\equiv \sum_{\varphi_{ij}} \prod_{i=1}^{N_b^x} \prod_{j=1}^{N_b^y} \hat{\mathcal{Z}}_{\mathrm{BG}}(\varphi_{i\pm1j}, \varphi_{ij}, \varphi_{ij\pm1}) \\
&= \sum_{\varphi_{ij}} \prod_{i=1}^{N_b^x} \prod_{j=1}^{N_b^y} \binom{\dim \mathbf{b}_{ij}}{\varphi_{ij} \dim \mathbf{b}_{ij}} \frac{\hat{\Psi}_{\mathrm{BG}}(\varphi_{i\pm1j}, \varphi_{ij}, \varphi_{ij\pm1}, \hat{\zeta}_{\mathrm{BG}}^{x,y}, \hat{\xi}_{\mathrm{BG}}^{x,y\pm})}{\hat{\Psi}_{\mathrm{BG}}(\varphi_{i\pm1j}, \varphi_{ij}, \varphi_{ij\pm1}, \zeta_{\mathrm{BG}}, \xi_{\mathrm{BG}}^{x,y\pm})} \times \\
&\quad \mathrm{e}^{-\dim \mathbf{b}_{ij}[\mathcal{H}_{\mathrm{BG}} - \tilde{h}(1-2\varphi_{ij}) - \mathcal{C}]},
\end{aligned}
\tag{3.21}
$$

where we have suppressed the arguments of $\mathcal{H}_{\mathrm{BG}}$, $\hat{\zeta}_{\mathrm{BG}}$, and $\hat{\xi}_{\mathrm{BG}}^{x,y\pm}$ for convenience. The BG Hamiltonian reads

$$
\begin{aligned}
&\mathcal{H}_{\mathrm{BG}}(\varphi_{i\pm1j}, \varphi_{ij}, \varphi_{ij\pm1}) \\
&= \bar{z}_x \tilde{J}_x (\hat{\zeta}_{\mathrm{BG}}^x(\varphi_{ij}, \varphi_{ij}) + (\hat{\xi}_{\mathrm{BG}}^{x+}(\varphi_{i+1j}, \varphi_{ij}) + \hat{\xi}_{\mathrm{BG}}^{x-}(\varphi_{i-1j}, \varphi_{ij}))/2) + (x \leftrightarrow y).
\end{aligned}
\tag{3.22}
$$

Finally, to determine the sum over φ_{ij} in Eq. (3.21), we again use the maximum term method. To that aim we introduce the BG free energy density in the thermodynamic limit of the spins

$$
\begin{aligned}
\tilde{\mathrm{f}}_{\mathrm{BG}}(\varphi_{i\pm1j}, \varphi_{ij}, \varphi_{ij\pm1}) &\equiv \lim_{\mathrm{s}}^{\mathbf{N}_\sigma} \left[-(\dim \mathbf{b}_{ij})^{-1} \ln (\hat{\mathcal{Z}}_{\mathrm{BG}}(\varphi_{i\pm1j}, \varphi_{ij}, \varphi_{ij\pm1})) \right] \\
&= \mathcal{H}_{\mathrm{BG}}(\varphi_{i\pm1j}, \varphi_{ij}, \varphi_{ij\pm1}) - \tilde{h}(1 - 2\varphi_{ij}) - \mathcal{C} \\
&+ (1 - 3\bar{z}/4)[\Xi(\varphi_{ij}) + \Xi(1 - \varphi_{ij})] \\
&+ (\bar{z}_x/4)[\Xi(1 - \varphi_{ij} - \hat{\zeta}_{\mathrm{BG}}^x) + \Xi(\varphi_{ij} - \hat{\zeta}_{\mathrm{BG}}^x) + 2\Xi(\hat{\zeta}_{\mathrm{BG}}^x)] \\
&+ (\bar{z}_x/8)\textstyle\sum_{\pm}[\Xi(1 - \varphi_{ij} - \hat{\xi}_{\mathrm{BG}}^{x\pm}) + \Xi(\varphi_{i\pm1j} - \hat{\xi}_{\mathrm{BG}}^{x\pm})] \\
&+ (\bar{z}_x/8)\textstyle\sum_{\pm}[\Xi(\varphi_{ij} - \varphi_{i\pm1j} + \hat{\xi}_{\mathrm{BG}}^{x\pm}) + \Xi(\hat{\xi}_{\mathrm{BG}}^{x\pm})] \\
&- (\bar{z}_x/8)\textstyle\sum_{\pm}[\Xi(1 - \varphi_{i\pm1j}) + \Xi(\varphi_{i\pm1j})] + (x \leftrightarrow y),
\end{aligned}
\tag{3.23}
$$

where the last term $(x \leftrightarrow y)$ applies to all terms with the prefactor $\bar{z}_x$, and we recall that $\Xi(x) \equiv x \ln(x)$. To optimize Eq. (3.23) over φ_{ij}, we can employ two different strategies:

1. Optimize $\tilde{\mathrm{f}}_{\mathrm{BG}}(\varphi_{i\pm1j}, \varphi_{ij}, \varphi_{ij\pm1})$ over φ_{ij} and finally apply $\lim_{\mathrm{s}}^{\mathbf{N}_b}[\cdot]$.
2. Apply $\lim_{\mathrm{s}}^{\mathbf{N}_b}[\tilde{\mathrm{f}}_{\mathrm{BG}}(\varphi_{i\pm1j}, \varphi_{ij}, \varphi_{ij\pm1})]$ and then optimize the free energy functional.

Below we carry out both strategies and show that they give equivalent results for the equilibrium concentration profile. Only the second strategy, however, leads to a Cahn-Hilliard type free energy functional.

3.6.1 Optimization w.r.t. $\varphi_{ij} \rightarrow \overset{N_b}{\underset{s}{\lim}}[\cdot]$

Extremizing Eq. (3.21) over φ_{ij} is equivalent to finding the local minima of the BG free density given by Eq. (3.23), yielding the equation

$$\sum_{k=\pm 1} \partial_{\varphi_{ij}} [\tilde{\mathrm{f}}_{\mathrm{BG}}(\varphi_{(i+k)\pm 1j}, \varphi_{(i+k)j}, \varphi_{(i+k)j\pm 1}) + \tilde{\mathrm{f}}_{\mathrm{BG}}(\varphi_{i\pm 1(j+k)}, \varphi_{i(j+k)}, \varphi_{i(j+k)\pm 1})]$$
$$+ \partial_{\varphi_{ij}} \tilde{\mathrm{f}}_{\mathrm{BG}}(\varphi_{i\pm 1j}, \varphi_{ij}, \varphi_{ij\pm 1}) \overset{!}{=} 0. \tag{3.24}$$

Upon taking the derivative of the BG free energy density w.r.t. φ_{ij}, we can use the following

$$\partial_{\hat{\zeta}^{x,y}_{\mathrm{BG}}} \tilde{\mathrm{f}}_{\mathrm{BG}}(\varphi_{i\pm 1j}, \varphi_{ij}, \varphi_{ij\pm 1}) = \partial_{\hat{\xi}^{x,y\pm}_{\mathrm{BG}}} \tilde{\mathrm{f}}_{\mathrm{BG}}(\varphi_{i\pm 1j}, \varphi_{ij}, \varphi_{ij\pm 1}) \overset{!}{=} 0, \tag{3.25}$$

since both $\hat{\zeta}^{x,y}_{\mathrm{BG}}$ and $\hat{\xi}^{x,y\pm}_{\mathrm{BG}}$ are derived by minimization of the BG free energy density. This renders the evaluation of Eq. (3.24) a relatively easy task and results in the following recurrent set of difference equations $\forall (i, j) \in (\{1, ..., N^x_b\}, \{1, ..., N^y_b\})$

$$\frac{\bar{z}_x}{8} \sum_{\pm} \left[\ln \left(\frac{1-\varphi_{ij}-\hat{\xi}^{x\pm}_{\mathrm{BG}}(\varphi_{i\pm 1j}, \varphi_{ij})}{\varphi_{ij}-\varphi_{i\pm 1j}+\hat{\xi}^{x\pm}_{\mathrm{BG}}(\varphi_{i\pm 1j}, \varphi_{ij})} \right) - \ln \left(\frac{\varphi_{ij}-\hat{\xi}^{x\pm}_{\mathrm{BG}}(\varphi_{ij}, \varphi_{i\pm 1j})}{\varphi_{i\pm 1j}-\varphi_{ij}+\hat{\xi}^{x\pm}_{\mathrm{BG}}(\varphi_{ij}, \varphi_{i\pm 1j})} \right) \right]$$
$$+ \frac{\bar{z}_y}{8} \sum_{\pm} \left[\ln \left(\frac{1-\varphi_{ij}-\hat{\xi}^{y\pm}_{\mathrm{BG}}(\varphi_{ij\pm 1}, \varphi_{ij})}{\varphi_{ij}-\varphi_{ij\pm 1}+\hat{\xi}^{y\pm}_{\mathrm{BG}}(\varphi_{ij\pm 1}, \varphi_{ij})} \right) - \ln \left(\frac{\varphi_{ij}-\hat{\xi}^{y\pm}_{\mathrm{BG}}(\varphi_{ij}, \varphi_{ij\pm 1})}{\varphi_{ij\pm 1}-\varphi_{ij}+\hat{\xi}^{y\pm}_{\mathrm{BG}}(\varphi_{ij}, \varphi_{ij\pm 1})} \right) \right]$$
$$= \frac{\bar{z}_x}{4} \ln \left(\frac{\varphi_{ij} - \hat{\zeta}^x_{\mathrm{BG}}(\varphi_{ij}, \varphi_{ij})}{1 - \varphi_{ij} - \hat{\zeta}^x_{\mathrm{BG}}(\varphi_{ij}, \varphi_{ij})} \right) + (x \leftrightarrow y) + (1 - \bar{z}) \ln \left(\frac{\varphi_{ij}}{1 - \varphi_{ij}} \right) + 2\tilde{h}. \tag{3.26}$$

For a one-dimensional concentration profile (i.e. $\varphi_{ij} \rightarrow \varphi_i$) a similar equation has been derived in [3]—see Eqs. (3.31)–(3.33) therein. Here we take the **thermodynamic limit of the spin blocks**, which transforms Eq. (3.26) into a PDE. To that aim we introduce the following notation:

$$\overset{\mathrm{N}_b}{\underset{\mathrm{s}}{\lim}} \left[\varphi_{ij} = \varphi(il_x, jl_y) \right] \equiv \varphi(x, y), \ \forall (x, y) \in A,$$
$$\overset{\mathrm{N}_b}{\underset{\mathrm{s}}{\lim}} \left[\varphi_{i\pm 1j} = \varphi(il_x \pm l_x, jl_y) \right] \equiv \lim_{l_x \rightarrow 0} \varphi(x \pm l_x, y), \ \forall (x, y) \in A,$$
$$\overset{\mathrm{N}_b}{\underset{\mathrm{s}}{\lim}} \left[\varphi_{i\pm 1j} = \varphi(il_x \pm l_x, jl_y) \right] \equiv \lim_{l_y \rightarrow 0} \varphi(x, y \pm l_y), \ \forall (x, y) \in A, \tag{3.27}$$

where $A = [-L_x/2, L_x/2] \times [-L_y/2, L_y/2]$. Applying Eq. (3.27) to the LHS of (3.26), we obtain the following limit for the numerators inside the first two logarithms:

$$\lim_{l_x \to 0} \Bigg[\sum_{\pm} \ln(1-\varphi(x,y)-\hat{\xi}^{x\pm}_{\text{BG}}(\varphi(x\pm l_x,y),\varphi(x,y)))- \\ 2\ln(1-\varphi(x,y)-\hat{\zeta}^{x}_{\text{BG}}(\varphi(x,y),\varphi(x,y)))\Bigg] / l_x^2 \\ = \frac{\hat{\xi}^{x(1,0)}_{\text{BG}}\partial_x^2\varphi(x,y) - \hat{\xi}^{x(2,0)}_{\text{BG}}(\partial_x\varphi(x,y))^2}{\varphi(x,y)+\hat{\zeta}^{x}_{\text{BG}}-1} - \left(\frac{\hat{\xi}^{x(1,0)}_{\text{BG}}\partial_x\varphi(x,y)}{\varphi(x,y)+\hat{\zeta}^{x}_{\text{BG}}-1}\right)^2, \tag{3.28}$$

$$\lim_{l_x \to 0} \Bigg[2\ln(\varphi(x,y)-\hat{\zeta}^{x}_{\text{BG}}(\varphi(x,y),\varphi(x,y))) \\ -\sum_{\pm} \ln(\varphi(x,y)-\hat{\xi}^{x\pm}_{\text{BG}}(\varphi(x,y),\varphi(x\pm l_x,y)))\Bigg] / l_x^2 \\ = \frac{\hat{\xi}^{x(0,1)}_{\text{BG}}\partial_x^2\varphi(x,y) + \hat{\xi}^{x(0,2)}_{\text{BG}}(\partial_x\varphi(x,y))^2}{\varphi(x,y)-\hat{\zeta}^{x}_{\text{BG}}} + \left(\frac{\hat{\xi}^{x(0,1)}_{\text{BG}}\partial_x\varphi(x,y)}{\varphi(x,y)-\hat{\zeta}^{x}_{\text{BG}}}\right)^2, \tag{3.29}$$

where $\hat{\xi}^{x(m,n)}_{\text{BG}} \equiv \partial_a^m \partial_b^n \hat{\xi}^{x}_{\text{BG}}(a,b)|_{(\varphi(x,y),\varphi(x,y))}$. The second term in (3.28) and first term in (3.29) are added manually, and therefore need to be added equally to the RHS of Eq. (3.26) by hand. Proceeding with the denominators inside the first two logarithms of (3.26), we get:

$$\lim_{l_x \to 0} \Bigg[\sum_{\pm} \ln(\varphi(x\pm l_x,y)-\varphi(x,y)+\hat{\xi}^{x\pm}_{\text{BG}}(\varphi(x,y),\varphi(x\pm l_x,y))) \\ -2\ln(\hat{\zeta}^{x}_{\text{BG}}(\varphi(x,y),\varphi(x,y)))\Bigg] / l_x^2 \\ = \frac{(\hat{\xi}^{x(0,1)}_{\text{BG}}+1)\partial_x^2\varphi(x,y) - \hat{\xi}^{x(0,2)}_{\text{BG}}(\partial_x\varphi(x,y))^2}{\hat{\zeta}^{x}_{\text{BG}}} - \left(\frac{(\hat{\xi}^{x(0,1)}_{\text{BG}}+1)\partial_x\varphi(x,y)}{\hat{\zeta}^{x}_{\text{BG}}}\right)^2, \tag{3.30}$$

$$\lim_{l_x \to 0} \Bigg[2\ln(\hat{\zeta}^{x}_{\text{BG}}(\varphi(x,y),\varphi(x,y))) \\ -\sum_{\pm} \ln(\varphi(x,y)-\varphi(x\pm l_x,y)+\hat{\xi}^{x\pm}_{\text{BG}}(\varphi(x\pm l_x,y),\varphi(x,y)))\Bigg] / l_x^2 \\ = \frac{(1-\hat{\xi}^{x(1,0)}_{\text{BG}})\partial_x^2\varphi(x,y) - \hat{\xi}^{x(2,0)}_{\text{BG}}(\partial_x\varphi(x,y))^2}{\hat{\zeta}^{x}_{\text{BG}}} + \left(\frac{(\hat{\xi}^{x(1,0)}_{\text{BG}}-1)\partial_x\varphi(x,y)}{\hat{\zeta}^{x}_{\text{BG}}}\right)^2. \tag{3.31}$$

Here the second term in (3.30) and first term in (3.31) are added manually. However, since these terms cancel each other, they shall not be added to the RHS of Eq. (3.26).

Upon interchanging $x \leftrightarrow y$ the results of Eqs. (3.28)–(3.31) readily apply to the y-direction. Summing up all the contributions, we obtain

$$\begin{aligned}&\frac{\bar{z}_x}{8}((3.28) + (3.29) + (3.30) + (3.31))\\&= \kappa_{\mathrm{BG},x}(\varphi(x, y))\partial_x^2\varphi(x, y) + \frac{\kappa'_{\mathrm{BG},x}(\varphi(x, y))(\partial_x\varphi(x, y))^2}{2},\end{aligned} \tag{3.32}$$

where $\kappa_{\mathrm{BG},x}(\varphi)$ is identified as the **gradient energy coefficient** given by

$$\kappa_{\mathrm{BG},(x,y)}(\varphi) \equiv \frac{\bar{z}_{x,y}(\mathrm{e}^{4\tilde{J}_{x,y}} - 1)}{4\sqrt{1 + 4(\mathrm{e}^{4\tilde{J}_{x,y}} - 1)\varphi(1 - \varphi)}}, \tag{3.33}$$

and $\kappa'_{\mathrm{BG},x}(\varphi) = \partial_\varphi\kappa_{\mathrm{BG},x}(\varphi)$. For a one-dimensional concentration profile the gradient energy coefficient has been derived in [4]—see Eq. (2.12b) therein—and agrees with our result. Plugging the result back into the LHS of Eq. (3.26), we obtain the following PDE (recall that $\hat{\zeta}_{\mathrm{BG}}^{x,y}$ is given by Eq. (3.17))

$$\begin{aligned}&l_x^2[\kappa_{\mathrm{BG},x}(\varphi(x, y))\partial_x^2\varphi(x, y) + \kappa'_{\mathrm{BG},x}(\varphi(x, y))(\partial_x\varphi(x, y))^2/2] + (x \leftrightarrow y)\\&= \frac{\bar{z}_x}{2}\ln\left(\frac{\varphi(x, y) - \hat{\zeta}_{\mathrm{BG}}^x}{1 - \varphi(x, y) - \hat{\zeta}_{\mathrm{BG}}^x}\right) + (x \leftrightarrow y) + (1 - \bar{z})\ln\left(\frac{\varphi(x, y)}{1 - \varphi(x, y)}\right) + 2\tilde{h},\end{aligned} \tag{3.34}$$

which applies in the domain $A = [-L_x/2, L_x/2] \times [-L_y/2, L_y/2]$.

3.6.2 $\lim_{\mathrm{s}}^{N_b}[\cdot] \rightarrow$ *Optimization w.r.t.* $\varphi(x, y)$

Here we proceed with the second strategy which results in a Cahn-Hilliard type free energy functional. Applying the thermodynamic limit of the blocks to Eq. (3.23) in the x-direction, we need to keep track of the following terms:

$$\begin{aligned}&\lim_{l_x \to 0}\left[\sum\nolimits_\pm \hat{\xi}_{\mathrm{BG}}^{x\pm}(\varphi(x \pm l_x, y), \varphi(x, y)) - 2\hat{\zeta}_{\mathrm{BG}}^x(\varphi(x, y), \varphi(x, y))\right]/l_x^2\\&\overset{\text{P.I.}}{=} [\partial_x\varphi(x, y)\hat{\xi}_{\mathrm{BG}}^{x(1,0)}]_{x=-L_x/2}^{x=L_x/2} - (\partial_x\varphi(x, y))^2\hat{\xi}_{\mathrm{BG}}^{x(1,1)},\end{aligned} \tag{3.35}$$

$$\begin{aligned}&\lim_{l_x \to 0}\left[\sum\nolimits_\pm \Xi(\varphi(x \pm l_x, y)) - 2\Xi(\varphi(x, y))\right]/l_x^2\\&\overset{\text{P.I.}}{=} [\partial_x\varphi(x, y)(\ln(\varphi(x, y)) + 1)]_{x=-L_x/2}^{x=L_x/2},\end{aligned} \tag{3.36}$$

$$\begin{aligned}&\lim_{l_x \to 0}\left[\sum\nolimits_\pm \Xi(1 - \varphi(x \pm l_x, y)) - 2\Xi(1 - \varphi(x, y))\right]/l_x^2\\&\overset{\text{P.I.}}{=} -[\partial_x\varphi(x, y)(\ln(1 - \varphi(x, y)) + 1)]_{x=-L_x/2}^{x=L_x/2},\end{aligned} \tag{3.37}$$

$$\lim_{l_x\to 0}\left[\sum\nolimits_{\pm}\Xi(\hat{\xi}^{x\pm}_{\rm BG}(\varphi(x\pm l_x,y),\varphi(x,y)))-2\Xi(\hat{\zeta}^{x}_{\rm BG}(\varphi(x,y),\varphi(x,y)))\right]/l_x^2$$
$$\overset{\rm P.I.}{=}[\partial_x\varphi(x,y)\hat{\xi}^{x(1,0)}_{\rm BG}(\ln(\hat{\zeta}^{x}_{\rm BG})+1)]^{x=L_x/2}_{x=-L_x/2}-(\partial_x\varphi(x,y))^2\hat{\xi}^{x(1,0)}_{\rm BG}\hat{\xi}^{x(0,1)}_{\rm BG}/\hat{\zeta}^{x}_{\rm BG}$$
$$-(\partial_x\varphi(x,y))^2\hat{\xi}^{x(1,1)}_{\rm BG}(\ln(\hat{\zeta}^{x}_{\rm BG})+1), \tag{3.38}$$

$$\lim_{l_x\to 0}\left[\sum\nolimits_{\pm}\Xi(\varphi(x\pm l_x,y)-\hat{\xi}^{x\pm}_{\rm BG}(\varphi(x\pm l_x,y),\varphi(x,y)))\right.$$
$$\left.-2\Xi(\varphi(x,y)-\hat{\zeta}^{x}_{\rm BG}(\varphi(x,y),\varphi(x,y)))\right]/l_x^2$$
$$\overset{\rm P.I.}{=}[\partial_x\varphi(x,y)(1-\hat{\xi}^{x(1,0)}_{\rm BG})(\ln(\varphi(x,y)-\hat{\zeta}^{x}_{\rm BG})+1)]^{x=L_x/2}_{x=-L_x/2}$$
$$-(\partial_x\varphi(x,y))^2\frac{(\hat{\xi}^{x(1,0)}_{\rm BG}-1)\hat{\xi}^{x(0,1)}_{\rm BG}}{\varphi(x,y)-\hat{\zeta}^{x}_{\rm BG}}+(\partial_x\varphi(x,y))^2\hat{\xi}^{x(1,1)}_{\rm BG}(\ln(\varphi(x,y)-\hat{\zeta}^{x}_{\rm BG})+1), \tag{3.39}$$

$$\lim_{l_x\to 0}\left[\sum\nolimits_{\pm}\Xi(1-\varphi(x,y)-\hat{\xi}^{x\pm}_{\rm BG}(\varphi(x\pm l_x,y),\varphi(x,y)))\right.$$
$$\left.-2\Xi(1-\varphi(x,y)-\hat{\zeta}^{x}_{\rm BG}(\varphi(x,y),\varphi(x,y)))\right]/l_x^2$$
$$\overset{\rm P.I.}{=}[-\partial_x\varphi(x,y)\hat{\xi}^{x(1,0)}_{\rm BG}(\ln(1-\varphi(x,y)-\hat{\zeta}^{x}_{\rm BG})+1)]^{x=L_x/2}_{x=-L_x/2}$$
$$-(\partial_x\varphi(x,y))^2\frac{\hat{\xi}^{x(1,0)}_{\rm BG}[\hat{\xi}^{x(0,1)}_{\rm BG}+1]}{1-\varphi(x,y)-\hat{\zeta}^{x}_{\rm BG}}$$
$$+(\partial_x\varphi(x,y))^2\hat{\xi}^{x(1,1)}_{\rm BG}(\ln(1-\varphi(x,y)-\hat{\zeta}^{x}_{\rm BG})+1), \tag{3.40}$$

$$\lim_{l_x\to 0}\left[\sum\nolimits_{\pm}\Xi(\varphi(x,y)-\varphi(x\pm l_x,y)+\hat{\xi}^{x\pm}_{\rm BG}(\varphi(x\pm l_x,y),\varphi(x,y)))\right.$$
$$\left.-2\Xi(\hat{\zeta}^{x}_{\rm BG}(\varphi(x,y),\varphi(x,y)))\right]/l_x^2$$
$$\overset{\rm P.I.}{=}[\partial_x\varphi(x,y)(\hat{\xi}^{x(1,0)}_{\rm BG}-1)(\ln(\hat{\zeta}^{x}_{\rm BG})+1)]^{x=L_x/2}_{x=-L_x/2}$$
$$-(\partial_x\varphi(x,y))^2\frac{(\hat{\xi}^{x(1,0)}_{\rm BG}-1)(\hat{\xi}^{x(0,1)}_{\rm BG}+1)}{\hat{\zeta}^{x}_{\rm BG}}-(\partial_x\varphi(x,y))^2\hat{\xi}^{x(1,1)}_{\rm BG}(\ln(\hat{\zeta}^{x}_{\rm BG})+1), \tag{3.41}$$

where we have immediately carried out partial integration—since each term arises inside an integral–and identify the first terms after the equal sign in Eqs. (3.35)–(3.41) with the **surface contribution**. Note that Eqs. (3.38)–(3.41) each contain three terms after the equal sign, which we respectively denote with $(3.38)_i$–$(3.41)_i$, $i=\{1,2,3\}$. We now add up the third terms $(3.38)_3$–$(3.41)_3$ which exactly cancel with Eq. (3.35) upon plugging them back into Eq. (3.23). Adding up the second terms $(3.38)_2$–$(3.41)_2$ yields

$$\frac{\bar{z}_x}{8}(3.38_2 + 3.39_2 + 3.40_2 + 3.41_2) = \frac{1}{2}\kappa_{\mathrm{BG},x}(\varphi(x,y))(\partial_x\varphi(x,y))^2. \tag{3.42}$$

Upon interchanging $x \leftrightarrow y$ the same results applies to the y-direction. Putting the results back into Eq. (3.23) and adding/subtracting those terms which have been added by hand in Eqs. (3.35)–(3.41), we finally obtain the BG free energy density

$$\lim_{\mathrm{s}}^{\mathbf{N}_b}\left[\tilde{\mathrm{f}}_{\mathrm{BG}}(\varphi_{i\pm1j},\varphi_{ij},\varphi_{ij\pm1})\right] = \tilde{\mathrm{f}}_{\mathrm{BG}}(\varphi(x,y)) + \frac{l_x^2}{2}\kappa_{\mathrm{BG},x}(\varphi(x,y))(\partial_x\varphi(x,y))^2 + (x \leftrightarrow y), \tag{3.43}$$

where $\kappa_{\mathrm{BG},x}(\varphi)$ is defined in Eq. (3.33), and the BG local free energy density with anisotropic coupling strength can be written as $\tilde{\mathrm{f}}_{\mathrm{BG}}(\varphi) = \tilde{\mathrm{u}}_{\mathrm{BG}}(\varphi) - \tilde{\mathrm{s}}_{\mathrm{BG}}(\varphi)$ with the internal energy and entropy density given by

$$\begin{aligned}\tilde{\mathrm{u}}_{\mathrm{BG}}(\varphi) &= 2[\bar{z}_x\tilde{J}_x(\hat{\zeta}^x_{\mathrm{BG}} - 1/8) + \bar{z}_y\tilde{J}_y(\hat{\zeta}^y_{\mathrm{BG}} - 1/8)] - \tilde{h}(1-2\varphi),\\ \tilde{\mathrm{s}}_{\mathrm{BG}}(\varphi) &= (\bar{z}-1)[\Xi(\varphi) + \Xi(1-\varphi)]\\ &\quad - \frac{\bar{z}_x}{2}[\Xi(\varphi - \hat{\zeta}^x_{\mathrm{BG}}) + \Xi(1-\varphi-\hat{\zeta}^x_{\mathrm{BG}}) + 2\Xi(\hat{\zeta}^x_{\mathrm{BG}})] + (x \leftrightarrow y),\end{aligned} \tag{3.44}$$

where $(x \leftrightarrow y)$ applies to the second term of $\tilde{\mathrm{s}}_{\mathrm{BG}}(\varphi)$. Note that for $\tilde{J}_x = \tilde{J}_y = \tilde{J}$ we get the isotropic free energy density given by Eq. (2.14). Finally, the BG free energy density functional—including the surface contribution—is given by

$$\begin{aligned}\tilde{F}_{\mathrm{BG}}[\varphi(x,y)] &\equiv \lim_{\mathrm{s}}^{\mathbf{N}_b}\left[(N_b^xN_b^y)^{-1}\sum_{i=1}^{N_b^x}\sum_{j=1}^{N_b^y}\tilde{\mathrm{f}}_{\mathrm{BG}}(\varphi_{i\pm1j},\varphi_{ij},\varphi_{ij\pm1})\right]\\ &= \frac{1}{l_xl_y}\int\limits_{(x,y)\in A}[\tilde{\mathrm{f}}_{\mathrm{BG}}(\varphi(x,y)) + \frac{l_x^2}{2}\kappa_{\mathrm{BG},x}(\partial_x\varphi(x,y))^2 + (x \leftrightarrow y)]dxdy\\ &\quad + \Phi_{\mathrm{BG}}[\varphi(x,y)],\end{aligned} \tag{3.45}$$

where the BG surface contribution reads

$$\Phi_{\mathrm{BG}}[\varphi(x,y)] = \frac{\bar{z}_xl_x}{8l_y}\int_y\left[\ln\left(\frac{2\exp(4\tilde{J}_x)(1-\varphi(x,y))}{\Omega_x(0,\zeta_{\mathrm{BG}}(\varphi(x,y)))-2\varphi(x,y)}\right)\partial_x\varphi(x,y)\right]_{x=-L_x/2}^{x=L_x/2}dy + (x \leftrightarrow y), \tag{3.46}$$

with ζ_{BG} and $\Omega_{x,y}(a,b)$ given by Eqs. (3.15) and (3.20), respectively. The profile $\varphi(x,y)$ which corresponds to a stationary point of Eq. (3.45), i.e. $\delta\tilde{F}_{\mathrm{BG}}[\varphi(x,y)]/\delta\varphi(x,y) = 0$, solves the corresponding E-L equation

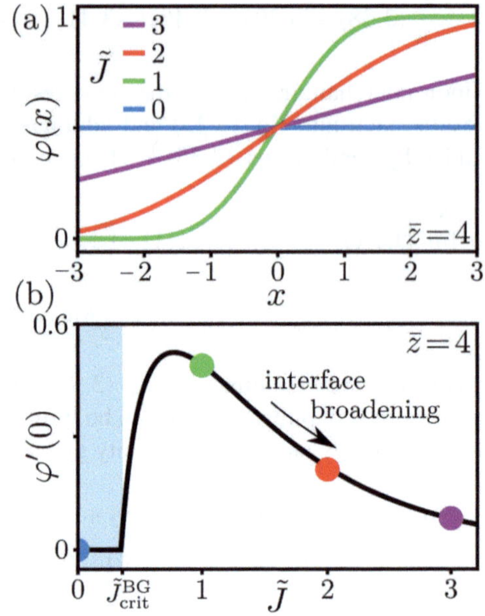

Fig. 3.2 Interface broadening in the BG one-dimensional concentration profile. a One-dimensional equilibrium profile $\varphi(x)$ that solves Eq. (3.47) for various values of $\tilde{J}$. **b** Interface steepness $\varphi'(0)$ as a function of $\tilde{J}$. Colored dots relate to the lines in (**a**). The blue shaded area corresponds to $\tilde{J} < \tilde{J}^{\mathrm{BG}}_{\mathrm{crit}}$ where the interface steepness remains zero

$$l_x^2[\kappa_{\mathrm{BG},x}\partial_x^2\varphi(x,y)+\kappa'_{\mathrm{BG},x}(\partial_x\varphi(x,y))^2/2]+(x\leftrightarrow y)=\partial_{\varphi(x,y)}\tilde{\mathrm{f}}_{\mathrm{BG}}(\varphi(x,y)), \quad (3.47)$$

with boundary conditions $\partial_{x,y}\varphi(x,y)|_{x,y=\pm L_{x,y}/2}=0$ and $\varphi(x,y)|_{x,y=\pm L_{x,y}/2}=\varphi^{\mathrm{BG}}_{\mathrm{min}}$ where $\varphi^{\mathrm{BG}}_{\mathrm{min}} \equiv \arg\inf_{0\leq\varphi\leq 1}\tilde{\mathrm{f}}_{\mathrm{BG}}(\varphi)$. Plugging Eq. (3.44) into Eq. (3.47) finally results in Eq. (3.34). This concludes our derivation of the Cahn-Hilliard free energy functional within the BG approximation.

3.7 Equilibrium Profile

Here we consider a concentration profile which only varies in the x direction, i.e. $\varphi(x,y)=\varphi(x)$, $\forall\, x\in[-L_x/2, L_x/2]$. In subsequent analysis x is defined in units of the block length l_x. The equilibrium profile $\varphi(x)$ is an extremum of Eq. (3.45) and solves Eq. (3.47). An example is given in Fig. 3.2a for a lattice with $\bar{z}=4$ obtained by numerically solving Eq. (3.47). Surprisingly, the interface steepness $\varphi'(0)\equiv d\varphi(x)/dx|_{x=0}$ is a non-monotonic function of $\tilde{J}$ as displayed in Fig. 3.2b. After a certain value around $\tilde{J}\approx 1$ the steepness decreases, corresponding to a broadening of the interface. In Chap. 4 we will give a physical explanation behind the **interface broadening**, whereas here we derive analytical expressions for the interface steepness, interface width (according to the Cahn-Hilliard definition [5]), and prove the broadening of the BG equilibrium profile for any lattice with $\bar{z}>2$.

3.7.1 Interface Steepness and Width

For a one-dimensional concentration profile Eq. (3.47) reduces to a second order autonomous ODE. To evaluate the interface steepness, we first rewrite the LHS of Eq. (3.47) as

$$\kappa_{\mathrm{BG},x}\varphi_{\mathrm{BG}}''(x) + \frac{1}{2}\kappa_{\mathrm{BG},x}'(\varphi_{\mathrm{BG}}'(x))^2 = \frac{1}{2\varphi_{\mathrm{BG}}'(x)}\frac{d}{dx}[\kappa_{\mathrm{BG},x}(\varphi_{\mathrm{BG}}'(x))^2]. \tag{3.48}$$

Taking the term $1/\varphi_{\mathrm{BG}}'(x)$ to the RHS of Eq. (3.47) and using the fact that $\varphi_{\mathrm{BG}}'(x)(\partial \mathrm{f}_{\mathrm{BG}}(\varphi_{\mathrm{BG}}(x))/\partial\varphi_{\mathrm{BG}}(x)) = d\mathrm{f}_{\mathrm{BG}}(\varphi_{\mathrm{BG}}(x))/dx$ by the chain rule, we can integrate both sides over x, resulting in the first-order autonomous ODE

$$\frac{1}{2}\kappa_{\mathrm{BG},x}(\varphi_{\mathrm{BG}}'(x))^2 = \tilde{\mathrm{f}}_{\mathrm{BG}}(\varphi_{\mathrm{BG}}(x)) + \mathcal{C}_1, \tag{3.49}$$

where $\mathcal{C}_1$ is an integration constant. From Eq. (3.49) we can directly readout the interface steepness

$$\varphi_{\mathrm{BG}}'(x) = \pm\sqrt{2(\tilde{\mathrm{f}}_{\mathrm{BG}}(\varphi_{\mathrm{BG}}(x)) - \tilde{\mathrm{f}}_{\mathrm{BG}}(\varphi_{\min}^{\mathrm{BG}}))/\kappa_{\mathrm{BG},x}(\varphi_{\mathrm{BG}}(x))}, \tag{3.50}$$

where the integration constant is set to $\mathcal{C}_1 = -\tilde{\mathrm{f}}_{\mathrm{BG}}(\varphi_{\min}^{\mathrm{BG}})$. The integration constant is chosen such that the term inside the square root on the RHS is always positive and to impose a vanishing derivative at the boundaries.

Now let us focus specifically on the isotropic case with a vanishing external field, i.e. $\tilde{J}_x = \tilde{J}_y = \tilde{J}$ and $\tilde{h} = 0$. Based on the analysis in Sect. 2.9, we know that the location of the global minimum can be written as $\varphi_{\min}^{\mathrm{BG}} = \chi_\varphi/(1+\chi_\varphi)$, where $\chi_\varphi \in [0,\infty)$ is given by the nontrivial solutions (i.e. $\chi_\varphi \neq 1$) of Eq. (2.27). Above the critical coupling $\tilde{J} > \tilde{J}_{\mathrm{crit}}^{\mathrm{BG}}$ there exists two nontrivial solutions, resulting in $\varphi_{\min}^{\mathrm{BG}} \neq 1/2$. For $\bar{z} = 4$ the result is given by Eq. (2.38), and for completeness we also give the results for $\bar{z} = 3$ and $\bar{z} = 6$, which reads

$$\begin{aligned}
\varphi_{\min}^{\mathrm{BG}}\Big|_{\tilde{h}=0}^{\bar{z}=3} &= \frac{1}{2}\left[1 \pm \frac{1}{2}\mathrm{Re}\left(\frac{\mathrm{e}^{2\tilde{J}}\sqrt{(\mathrm{e}^{2\tilde{J}}+1)(\mathrm{e}^{2\tilde{J}}-3)}}{\mathrm{e}^{3\tilde{J}}\sinh(\tilde{J})-1}\right)\right],\\
\varphi_{\min}^{\mathrm{BG}}\Big|_{\tilde{h}=0}^{\bar{z}=6} &= \mathrm{Re}\left(\frac{(\mathrm{e}^{2\tilde{J}}+(\mathrm{e}^{4\tilde{J}}+4)^{\frac{1}{2}} \pm \sqrt{2}(\mathrm{e}^{2\tilde{J}}(\mathrm{e}^{4\tilde{J}}+4)^{\frac{1}{2}}+\mathrm{e}^{4\tilde{J}}-6)^{\frac{1}{2}})^6}{4096+(\mathrm{e}^{2\tilde{J}}+(\mathrm{e}^{4\tilde{J}}+4)^{\frac{1}{2}}+\sqrt{2}(\mathrm{e}^{2\tilde{J}}(\mathrm{e}^{4\tilde{J}}+4)^{\frac{1}{2}}+\mathrm{e}^{4\tilde{J}}-6)^{\frac{1}{2}})^6}\right),
\end{aligned} \tag{3.51}$$

where $\mathrm{Re}(\cdot)$ denotes the real part. For $\tilde{h} \neq 0$ Eq. (2.27) is also explicitly solvable for $\bar{z} = \{3, 4\}$ but the final expression is less compact (see for example Eqs. (2.30a) and (2.30b) for $\bar{z} = 4$). Plugging the expression for $\varphi_{\min}^{\mathrm{BG}}$ into Eq. (3.50) and noting that $\varphi_{\mathrm{BG}}(0) = 1/2$, we have closed-form expressions for the interface steepness at

$x = 0$. To get the interface width as defined by Cahn and Hilliard—see Eq. (2.25) in [5]—we simply take a line tangential to the slope of the concentration profile at $x = 0$ and determine the crossing points of this line with the bulk concentration values as depicted in Fig. 3.3a. This leads to the expression

$$l_{\mathrm{BG,CH}} = |2\varphi_{\mathrm{min}}^{\mathrm{BG}} - 1|/|\varphi'_{\mathrm{BG}}(0)|. \tag{3.52}$$

Similarly to the interface steepness, we see in Fig. 3.3b that the BG interface width (blue line) is a non-monotonic function of $\tilde{J}$. Beyond a certain threshold value the interface width increases, resulting in a broadening of the interface.

3.7.2 *Interface Broadening for Infinite Coupling*

To prove that broadening is a general effect regardless of the lattice we take the strong coupling limit of Eq. (3.50). For $\bar{z} > 2$ and $\tilde{J} \to \infty$ the nontrivial solutions to Eq. (2.27) are approaching $\chi_\varphi \to 0$ and $\chi_\varphi \to \infty$, resulting in $\varphi_{\mathrm{min}}^{\mathrm{BG}} \to 0 \vee 1$. Plugging this into Eq. (3.50) together with $\varphi_{\mathrm{BG}}(0) = 1/2$, we obtain

$$\lim_{\tilde{J}\to\infty} \varphi'_{\mathrm{BG}}(0) = \lim_{\tilde{J}\to\infty} \pm\sqrt{2(2\bar{z}\tilde{J} - \bar{z}\ln(\mathrm{e}^{2\tilde{J}} + 1) + (\bar{z} - 2)\ln(2))/\bar{z}_x \sinh(2\tilde{J})} = 0. \tag{3.53}$$

So we find a vanishing interface steepness at $x = 0$ for any lattice with $\bar{z} > 2$ in the strong coupling limit. Similarly for the interface width, we find

$$\lim_{\tilde{J}\to\infty} l_{\mathrm{BG,CH}} = \lim_{\tilde{J}\to\infty} |2\varphi_{\mathrm{min}}^{\mathrm{BG}} - 1|/\varphi'_{\mathrm{BG}}(0) = \infty. \tag{3.54}$$

Hence, in the strong-interaction limit the interface width diverges for any lattice with $\bar{z} > 2$.

3.8 Linear Stability Analysis

Here we determine the length scales on which inhomogeneities of the concentration profile are stable. We consider a concentration profile of the form $\varphi(\mathbf{x}) = \varphi_0 + a\sin(\mathbf{q}\cdot\mathbf{x})$ with $\mathbf{q} = (q_x, q_y)^{\mathrm{T}}$ and $|a| \ll \min(\varphi_0, 1 - \varphi_0)$. A sinusoidal perturbation is taken to agree with the odd boundary conditions which we imposed for Eq. (3.47). Expanding the local free energy density and gradient energy coefficient around the homogeneous state up to second order gives

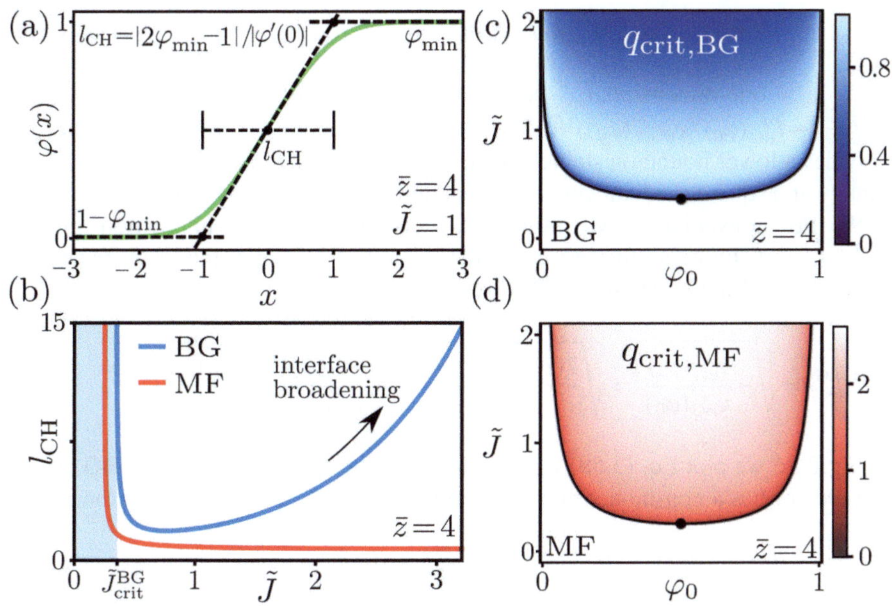

Fig. 3.3 **a** Example of the Cahn-Hilliard interface width l_{CH} used in Eq. (3.52). **b** Cahn-Hilliard interface width within the BG (blue) and MF (red) approximation. **c–d** Critical wavevector within the BG (3.59) (**c**) and MF (3.72) (**d**) approximation. Brighter colors correspond to larger values for q_{crit}. The black line represents the spinodal given by (2.25) and (2.41), respectively. The black dot shows the BG and MF critical point

$$\tilde{\text{f}}_{\text{BG}}(\varphi(\mathbf{x})) = \tilde{\text{f}}_{\text{BG}}(\varphi_0) + a\sin(\mathbf{q}\cdot\mathbf{x})\tilde{\text{f}}'_{\text{BG}}(\varphi_0) + \frac{1}{2}a^2\sin^2(\mathbf{q}\cdot\mathbf{x})\tilde{\text{f}}''_{\text{BG}}(\varphi_0) + \mathcal{O}(a^3), \tag{3.55}$$

$$\frac{1}{2}\nabla\varphi(\mathbf{x})^{\text{T}}\boldsymbol{\kappa}_{\text{BG}}(\varphi(\mathbf{x}))\nabla\varphi(\mathbf{x}) = \frac{1}{2}a^2(\mathbf{q}^{\text{T}}\boldsymbol{\kappa}_{\text{BG}}(\varphi_0)\mathbf{q})\cos^2(\mathbf{q}\cdot\mathbf{x}) + \mathcal{O}(a^3), \tag{3.56}$$

where $\tilde{\text{f}}'_{\text{BG}}(\varphi_0) \equiv \partial_\varphi\tilde{\text{f}}_{\text{BG}}(\varphi)|_{\varphi_0}$ and $\tilde{\text{f}}''_{\text{BG}}(\varphi_0) \equiv \partial^2_\varphi\tilde{\text{f}}_{\text{BG}}(\varphi)|_{\varphi_0}$. Now we want to find out when a sinusoidal perturbation decreases the total free energy compared to the uniform concentration profile. Plugging Eqs. (3.55) and (3.56) into Eq. (3.45) and subtracting the free energy density of the uniform concentration gives

$$\tilde{F}_{\text{BG}}[\varphi(\mathbf{x})] - \tilde{F}_{\text{BG}}[\varphi_0] = \frac{a^2}{4}\left(\tilde{\text{f}}''_{\text{BG}}(\varphi_0) + \mathbf{q}^{\text{T}}\boldsymbol{\kappa}_{\text{BG}}(\varphi_0)\mathbf{q}\right) + \mathcal{O}\left(\frac{a^2}{L_xL_y}\right), \tag{3.57}$$

where we have taken the large system-size limit $(L_x, L_y) \to \infty$. To decrease the total free energy, the RHS of Eq. (3.57) must be negative. Note that $\mathbf{q}^{\text{T}}\boldsymbol{\kappa}_{\text{BG}}(\varphi_0)\mathbf{q} \geq 0$, and therefore only $\tilde{\text{f}}''_{\text{BG}}(\varphi_0)$ can make the RHS negative. The region where $\tilde{\text{f}}''_{\text{BG}}(\varphi_0) < 0$ in the $(\varphi_0, \tilde{J})$-plane is called the spinodal region (see also Sect. 2.8.2), and therefore this process is also known as **spinodal decomposition**. When $\tilde{\text{f}}''_{\text{BG}}(\varphi_0) < 0$ there is an upper bound on stable wavevectors which is given by

$$\mathbf{q}_{\rm crit}^{\rm T}\boldsymbol{\kappa}_{\rm BG}(\varphi_0)\mathbf{q}_{\rm crit} = -\tilde{\rm f}''_{\rm BG}(\varphi_0). \tag{3.58}$$

For a one-dimensional perturbation ($q_y = 0$) this translates to $q_{\rm crit}^{\rm BG} = (-\tilde{\rm f}''_{\rm BG}(\varphi_0)/\kappa_{{\rm BG},x}(\varphi_0))^{1/2}$. The critical wavelength given by $\lambda_{\rm crit}^{\rm BG} = 2\pi/q_{\rm crit}^{\rm BG}$ provides a lower bound on stable wavelengths. Recall that the BG local free energy density and square gradient coefficient are defined in Eqs. (3.44) and (3.33), respectively. For convenience, we immediately take the isotropic interaction strength $\tilde{J}_x = \tilde{J}_y = \tilde{J}$. Plugging the results for the second derivative of the local free energy density—see Eq. (2.23)—into $q_{\rm crit}^{\rm BG}$ gives

$$q_{\rm crit}^{\rm BG} = \sqrt{-\frac{\tilde{\rm f}''_{\rm BG}(\varphi_0)}{\kappa_{\rm BG}(\varphi_0)}} = \sqrt{\frac{2(\bar{z}-2)(1+4(e^{4\tilde{J}}-1)\varphi_0(1-\varphi_0))^{\frac{1}{2}} - 2\bar{z}}{\bar{z}_x\varphi_0(1-\varphi_0)(e^{4\tilde{J}}-1)}}. \tag{3.59}$$

In Fig. 3.3c we plot Eq. (3.59) for $\bar{z} = 4$. Inside the spinodal region the BG critical wavevector has a non-monotonic trend and for $0 < \varphi_0 < 1$ converges to the value

$$\lim_{\tilde{J}\to\infty} q_{\rm crit}^{\rm BG} = 0. \tag{3.60}$$

Similarly, the critical wavelength diverges, i.e. $\lim_{\tilde{J}\to\infty}\lambda_{\rm crit}^{\rm BG} = \infty$. Hence for $0 < \varphi_0 < 1$ there exist no finite stable wavelength perturbations in the strong interaction limit. The coupling strength $\tilde{J}^\dagger(\varphi_0)$ where $q_{\rm crit}^{\rm BG}$ is maximal—and therefore $\lambda_{\rm crit}^{\rm BG}$ minimal—reads

$$\tilde{J}^\dagger(\varphi_0) = \frac{1}{4}\ln\left(1 + \frac{\bar{z}(2+\sqrt{\bar{z}-1})-2}{(\bar{z}-2)^2\varphi_0(1-\varphi_0)}\right), \tag{3.61}$$

with the corresponding $\lambda_{\rm crit}^{\rm BG}(\tilde{J}^\dagger(\varphi_0))$ given by

$$\lambda_{\rm crit}^{\rm BG}(\tilde{J}^\dagger(\varphi_0)) = \frac{\pi\sqrt{\bar{z}_x}}{|\bar{z}-2|}\sqrt{\frac{\bar{z}(2+\sqrt{\bar{z}-1})-2}{(\bar{z}(1+\sqrt{\bar{z}-1}+\bar{z}/4)-1)^{\frac{1}{2}} - \bar{z}/2}}, \tag{3.62}$$

which is independent of the uniform background concentration $0 < \varphi_0 < 1$. In Chap. 4 we will further elaborate on the non-monotonic nature of the critical wavevector and wavelength in the BG approximation.

3.9 Mean Field Approximation

Here we list the results obtained with the MF approximation for sake of completeness. Similarly to the results in Chap. 2, the BG results converge to the MF results for $\tilde{J}_{x,y} \to 0$ or $\bar{z} \to \infty$.

3.9.1 Partition Function

On the MF level we introduce the following approximation for the fraction of defects between two blocks b_{ij} and b_{mn}:

$$\zeta_{\mathrm{MF}}(\varphi_{ij}, \varphi_{mn}) \equiv (\varphi_{ij}(1-\varphi_{mn}) + \varphi_{mn}(1-\varphi_{ij}))/2. \tag{3.63}$$

We thus approximate the number of defects between blocks b_{ij} and b_{mn} by the product of the *spin down* concentration in box b_{ij} and *spin up* concentration in box b_{mn}, and vice versa. This results in the following MF partition function

$$\mathcal{Z}_{\mathrm{MF}} = \sum_{\varphi_{ij}} \prod_{i=1}^{N_b^x} \prod_{j=1}^{N_b^y} \binom{\dim \mathbf{b}_{ij}}{\varphi_{ij} \dim \mathbf{b}_{ij}} \mathrm{e}^{-\dim \mathbf{b}_{ij}[\tilde{\mathcal{H}}_{\mathrm{MF}}(\varphi_{i\pm1j}, \varphi_{ij}, \varphi_{ij\pm1}) - \tilde{h}(1-2\varphi_{ij}) - \mathcal{C}]}, \tag{3.64}$$

where the MF Hamiltonian reads (omitting the arguments for convenience)

$$\tilde{\mathcal{H}}_{\mathrm{MF}} = \bar{z}_x \tilde{J}_x[\zeta_{\mathrm{MF}}(\varphi_{ij}, \varphi_{ij}) + (\zeta_{\mathrm{MF}}(\varphi_{i+1j}, \varphi_{ij}) + \zeta_{\mathrm{MF}}(\varphi_{i-1j}, \varphi_{ij}))/2] + (x \leftrightarrow y). \tag{3.65}$$

3.9.2 Free Energy Functional

The MF free energy density functional is given by

$$\tilde{F}_{\mathrm{MF}}[\varphi(x,y)] = \frac{1}{l_x l_y} \int\limits_{(x,y)\in A} [\tilde{\mathrm{f}}_{\mathrm{MF}}(\varphi(x,y)) + \frac{l_x^2}{2}\kappa_{\mathrm{MF},x}(\partial_x \varphi(x,y))^2 + (x \leftrightarrow y)]dxdy$$
$$+\Phi_{\mathrm{MF}}[\varphi(x,y)], \tag{3.66}$$

where the gradient energy coefficient is given by $\kappa_{\mathrm{MF},(x,y)} \equiv \bar{z}_{x,y}\tilde{J}_{x,y}$ which is independent of $\varphi(x,y)$. The MF surface contribution is given by

$$\Phi_{\mathrm{MF}}[\varphi(x,y)] = \frac{\bar{z}_x \tilde{J}_x l_x}{2 l_y} \int\limits_y [(1-\varphi(x,y))\partial_x \varphi(x,y)]_{x=-L_x/2}^{x=L_x/2} dy + (x \leftrightarrow y), \tag{3.67}$$

and finally the local free energy density entering in Eq. (3.66) reads

$$\tilde{\mathrm{f}}_{\mathrm{MF}}(\varphi) \equiv 2(\bar{z}_x \tilde{J}_x + \bar{z}_y \tilde{J}_y)[\varphi(1-\varphi) - 1/4] - \tilde{h}(1-2\varphi) + \Xi(\varphi) + \Xi(1-\varphi). \tag{3.68}$$

Note that for $\tilde{J}_x = \tilde{J}_y = \tilde{J}$ Eq. (3.68) is equal to Eq. (2.40).

3.9.3 Interface Steepness and Width

The MF interface steepness for a one-dimensional profile $\varphi(x)$ reads

$$\varphi'_{\mathrm{MF}}(0) = \pm\sqrt{2(\tilde{\mathrm{f}}_{\mathrm{MF}}(1/2) - \tilde{\mathrm{f}}_{\mathrm{MF}}(\varphi_{\min}^{\mathrm{MF}}))/\bar{z}_x \tilde{J}_x}, \tag{3.69}$$

where $\varphi_{\min}^{\mathrm{MF}} \equiv \arg\inf_{0\le\varphi\le} \tilde{\mathrm{f}}_{\mathrm{MF}}(\varphi)$ is the global minimum of the free energy density given by Eq. (2.42). The MF Cahn-Hilliard interface width can directly be obtained from Eq. (3.52) upon substituting in the MF expressions. In Fig. 3.3b we see that the MF interface width (red line) decreases monotonically with $\tilde{J}$ and does *not* broaden.

3.9.4 Absence of Interface Broadening

Let us consider an isotropic coupling $\tilde{J}_x = \tilde{J}_y = \tilde{J}$. In the limit of infinite coupling strength the solutions to Eq. (2.42) are given by $\varphi_{\min}^{\mathrm{MF}} \to 0 \vee 1,\ \forall \bar{z} > 0$, and therefore we get

$$\lim_{\tilde{J}\to\infty} \varphi'_{\mathrm{MF}}(0) = \lim_{\tilde{J}\to\infty} \pm\sqrt{2(\bar{z}\tilde{J}/2 - \ln(2))/\bar{z}_x \tilde{J}} = \pm\sqrt{\bar{z}/\bar{z}_x}. \tag{3.70}$$

Hence, in the infinite coupling limit the interface steepness converges to a maximum finite nonzero value. Furthermore, the interface width decreases and converges to the value

$$\lim_{\tilde{J}\to\infty} l_{\mathrm{MF,CH}} = \lim_{\tilde{J}\to\infty} |2\varphi_{\min}^{\mathrm{MF}} - 1|/|\varphi'_{\mathrm{MF}}(0)| = \sqrt{\bar{z}_x/\bar{z}}. \tag{3.71}$$

3.9.5 Spinodal Decomposition

Taking the MF local free energy density and square gradient coefficient and plugging them into q_{crit} defined in Eq. (3.58) gives

$$q_{\mathrm{crit}}^{\mathrm{MF}} = \sqrt{-\frac{\tilde{\mathrm{f}}''_{\mathrm{MF}}(\varphi_0)}{\kappa_{\mathrm{MF}}}} = \sqrt{\frac{4(\bar{z}_x \tilde{J}_x + \bar{z}_y \tilde{J}_y) - 1/(\varphi_0(1-\varphi_0))}{\bar{z}_x \tilde{J}_x}}. \tag{3.72}$$

For isotropic interaction strength $\tilde{J}_x = \tilde{J}_y = \tilde{J}$ and inside the spinodal region the MF critical wavevector is monotonically increasing with $\tilde{J}$ and for $0 < \varphi_0 < 1$ converges to

$$\lim_{\tilde{J}\to\infty} q_{\mathrm{crit}}^{\mathrm{MF}} = 2\sqrt{\bar{z}/\bar{z}_x}. \tag{3.73}$$

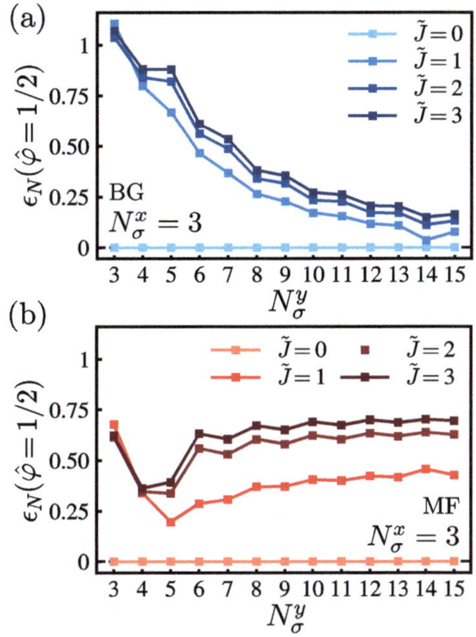

Fig. 3.4 Relative error between the exact and approximated partition function obtained with the **a** MF and **b** BG approximation for increasing number of spins and various values of the coupling strength $\tilde{J} = \{0, 1, 2, 3\}$. The relative error in Eq. (3.75) is determined for a square lattice composed of $(N_\sigma^x = 3) \times (N_\sigma^y = \{3, ..., 15\})$ spins with periodic boundary conditions in the vertical and fixed boundary conditions in the horizontal direction, respectively. The total fraction of down spins is fixed to $\hat{\varphi}=1/2$

In Fig. 3.3d we plot Eq. (3.72) for $\bar{z} = 4$ with isotropic interaction strength. The critical wavevector increases monotonically with $\tilde{J}$ and converges to the aforementioned result.

3.10 Error Analysis

To probe the accuracy of the MF and BG approximations, we compare their partition functions given by Eqs. (3.64) and (3.21) with exact results for the partition function of finite systems. For a uniform system an error analysis between the MF and BG approximation is given in Sect. 2.11.2, and here we proceed with the non-uniform analysis. We limit our error analysis to a one-dimensional concentration profile, conform with the majority of results discussed in this chapter. For a lattice composed of $N_\sigma^x \times N_\sigma^y$ spins, let $\boldsymbol{\varphi} = (\varphi_1, ..., \varphi_{N_\sigma^x})$ be a vector containing the concentration of downs spins in each column of the lattice. The total concentration of down spins in the lattice is given by $\varphi = ||\boldsymbol{\varphi}||_1/N_\sigma^x$. The exact partition function for a fixed concentration profile along the columns is denoted with $\mathcal{Z}(\boldsymbol{\varphi})$ and can be computed via

$$\mathcal{Z}(\hat{\boldsymbol{\varphi}}) = \sum_{\boldsymbol{\sigma}} \mathrm{e}^{-\tilde{\mathcal{H}}(\boldsymbol{\sigma})} \prod_{i=1}^{N_\sigma^x} \mathbb{1}_{\hat{\varphi}_i} [\varphi_i] \,, \tag{3.74}$$

where we recall that $\boldsymbol{\sigma}$ denotes the matrix containing all spin configurations, $\mathbb{1}_x[z]$ is the indicator function of x, and $\mathcal{H}(\boldsymbol{\sigma})$ is the Ising Hamiltonian. The relative error between $\mathcal{Z}_{\mathrm{BG,MF}}(\boldsymbol{\varphi})$ and $\mathcal{Z}(\boldsymbol{\varphi})$ for a fixed total concentration of down spins $\hat{\varphi}$ is defined as

$$\epsilon_N(\hat{\varphi}) = \left(\sum_{\boldsymbol{\varphi}} \mathcal{Z}(\boldsymbol{\varphi}) \left(1 - \frac{\ln\left(\mathcal{Z}_{\mathrm{MF,BG}}(\boldsymbol{\varphi})\right)}{\ln\left(\mathcal{Z}(\boldsymbol{\varphi})\right)}\right) \mathbb{1}_{\hat{\varphi}}[\varphi]\right) / \left(\sum_{\boldsymbol{\varphi}} \mathcal{Z}(\boldsymbol{\varphi}) \mathbb{1}_{\hat{\varphi}}[\varphi]\right). \tag{3.75}$$

Equation (3.75) is defined such that differences between $\mathcal{Z}(\boldsymbol{\varphi})$ and $\mathcal{Z}_{\mathrm{MF,BF}}(\boldsymbol{\varphi})$ attain the largest weight for thermodynamically stable configurations. In Fig. 3.4 we plot the relative error for the (a) MF and (b) BG approximation for a finite square lattice composed of $(N_\sigma^x = 3) \times (N_\sigma^y = \{3, ..., 15\})$ spins with fixed and periodic boundary conditions in the horizontal and vertical direction, respectively. Upon increasing the number of spins in the vertical direction, we see that the relative error of the BG approximation decreases towards zero, regardless of the coupling strength, whereas the MF approximation saturates to a nonzero value (note that the small system size gives rise to a marked even-odd dependency). For $\tilde{J} = 0$ both approximations are exact and therefore have zero relative error. The improvement of the BG approximation with increasing N_σ^y is due to the fact that Eq. (3.21) is obtained through a variational principle which is applied in the thermodynamic scaling limit. The MF approximation on the other hand becomes worse with increasing N_σ^y due to the approximation for the fraction of defects given by Eq. (3.63).

References

1. Blom K, Ziethen N, Godec A (2022) Thermodynamically consistent phase-field theory including nearest-neighbor pair correlations. Phys Rev Res 5(1):013135. [Online]. Available https://link.aps.org/doi/10.1103/PhysRevResearch.5.013135
2. Kadanoff LP (1966) Scaling laws for Ising models near T_c. Phys Physique Fizika 2:263–272. [Online]. Available https://link.aps.org/doi/10.1103/PhysicsPhysiqueFizika.2.263
3. Parlange J (1968) Phase transition and surface tension in the Guasichemical approximation. J Chem Phys 48(1):169–173. [Online]. Available https://doi.org/10.1063/1.1667895
4. Kikuchi R, Cahn JW (1962) Theory of domain walls in ordered structures—II: pair approximation for nonzero temperatures. J Phys Chem Solids 23(1-2):137–151. [Online]. Available https://doi.org/10.1016/0022-3697(62)90067-7
5. Cahn JW, Hilliard JE (1958) Free energy of a nonuniform system. I. Interfacial free energy. J Chem Phys 28(2):258–267. [Online]. Available https://aip.scitation.org/doi/10.1063/1.1744102

Chapter 4
Delocalization-Induced Interface Broadening in Strongly Interacting Systems

When you are studying any matter, or considering any philosophy, ask yourself only "What are the facts, and what is the truth that the facts bear out?".

Bertrand Russell, Message to Future Generations (1959)

In this chapter we illuminate unexpected effects of correlations on the structure and thermodynamics of interfaces and in turn phase separation, which are decisive in systems with strong interactions. More specifically, in Chap. 3 we encountered an unexpected effect of the BG concentration profile, which we named **interface broadening**. Based on the derived Cahn-Hilliard field theory, we found that the one-dimensional equilibrium concentration profile broadens upon increasing the coupling strength[1] above and near the thermal energy (see Fig. 3.2). Thus far we have not explained *what the underlying physical mechanism behind interface broadening is*. In this chapter we show that interface broadening is the result of an **entropy-driven interface delocalization transition**, which is not accounted for in the widely adopted MF theory. Pair correlations enforce a thermodynamically optimal configuration of defects, and profoundly affect nucleation and spinodal decomposition at strong coupling. Part of this chapter has been published in *Physical Review Research* under the terms of the Creative Commons Attribution 4.0 International license [1].

[1] Or similarly decreasing the temperature.

K. Blom, *Pair-Correlation Effects in Many-Body Systems*, Springer Theses,
https://doi.org/10.1007/978-3-031-29612-3_4

Fig. 4.1 **Capillary wave fluctuations** (red) along an Ising interface

4.1 Introduction and Motivation

Instigated by the seminal works of Cahn and Hilliard [2–4], phase separation—the process through which distinct phases form from a homogeneous mixture—has attracted considerable attention in a variety of fields, incl. physics [5–14], mathematics [15–17], chemistry [18–21], material science [22–24], and recently biology [25–28]. Our basic understanding of phase separation in systems in [29, 30] and out [31, 32] of equilibrium is mostly based on MF ideas [33], also known as regular solution [2], Bragg-Williams [34] (see Sect. 1.1.2), or Flory-Huggins [35, 36] theory (for recent works see [11–14, 26–28, 37–42]). MF theory neglects correlations whose importance grows with the strength of interactions [5] (see for example Sect. 2.11). Furthermore, **capillary wave fluctuations**[2] [42, 43] (see Fig. 4.1) and **interface delocalization** [44–55] are *not* captured in MF theories.

Various refined techniques have been developed beyond the MF approximation, incl. the cavity method [56], random phase approximation [57, 58], self-consistent field theory [59], and field-theoretic approaches close to criticality [60]. Yet, these techniques do not apply to non-uniform systems or are applicable in a limited range of interaction strengths. As a result, the phenomenology of phase separation in the strong-coupling limit remains largely unexplored and thus poorly understood.

Here we employ the BG approximation [61–64] that takes into account nearest-neighbor pair-correlations. In Chap. 3 we derived a Cahn-Hilliard field theory for a spatially inhomogeneous two-dimensional nearest-neighbor Ising model, based on the BG approximation. In this chapter we show that the derived field theory effectively accounts for the effects of capillary wave fluctuations. Upon a decrease in the capillary wave amplitude at sufficiently strong interactions, the interface *effectively broadens* due to **translation invariance** of the interface position. We corroborate the broadening with computer simulations and exact results in the infinite-interaction limit. Furthermore, via numerical simulations of the **Cahn-Hilliard equation** [65],

[2] Capillary waves are **long-wavelength fluctuations** of the interface profile (see Fig. 4.1).

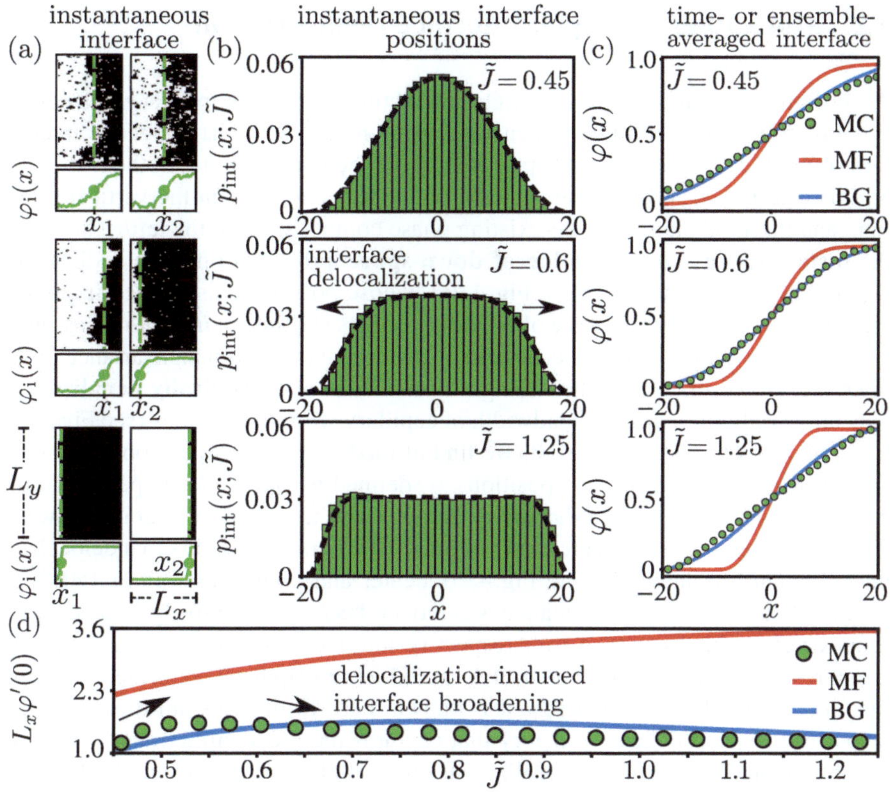

Fig. 4.2 **a** Realizations of spin configurations (top) and corresponding instantaneous interfaces (bottom) in a 2D Ising strip with lattice constant Δ and dimensions $(L_x, L_y) = (40, 120)\Delta$ for different $\tilde{J}$ values obtained from MC simulations (see Appendix 1 for simulation details); the circle and dashed line denote the position of the instantaneous interface. **b** Statistics of interface positions derived from simulations (green) and given by Eq. (4.16) in Appendix 2 (line) via a mapping onto a **Brownian excursion problem**. **c** Corresponding ensemble averaged concentration profile along the x-axis alongside theoretical predictions of MF (red) and BG (blue) theory. **d** Scaled interface steepness $L_x\varphi'(0) \equiv L_x\partial_x\varphi(x)|_{x=0}$ as a function of $\tilde{J}$

we analyze nucleation kinetics, and observe amplified nucleation barriers and a non-monotonic dependence of the interface steepness and critical nucleus size on the interaction strength.

4.2 Motivating Example: Interface Delocalization

An intriguing phenomenon in strongly interacting systems is interface delocalization [44–55]. Consider a two-dimensional Ising model with ferromagnetic interaction strength $\tilde{J} = J/k_{\mathrm{B}}T$ in a strip geometry (i.e. height $\gg$ length) in the two-phase regime (see Fig. 4.1). Imposing periodic boundary conditions in the vertical direction, and thermodynamically co-existing phase compositions at the left/right edges, the **instantaneous concentration of down-spins**[3] projected onto the x direction, $\varphi_{\mathrm{i}}(x)$, develops an interface (see Fig. 4.2a), whose position x_{i} is defined implicitly via $\varphi_{\mathrm{i}}(x_{\mathrm{i}}) = 1/2$. In the absence of boundary effects shifting an *instantaneous* interface $\varphi_{\mathrm{i}}(x_{\mathrm{i}}) \to \varphi_{\mathrm{i}}(x_{\mathrm{i}} + \mathrm{d}x_{\mathrm{i}})$ costs no energy, rendering translational invariance of the interface. However, x_{i} near the left/right boundaries are **entropically penalized**, as they allow only for a limited bandwidth of capillary wave fluctuations (see Fig. 4.2a, top) [54, 55, 66, 67]. As a result, we find at moderate $\tilde{J}$ that the probability density of instantaneous interface positions x_{i}, defined as $p_{\mathrm{int}}(x; \tilde{J})$, is peaked at the center (see Fig. 4.2b, top), whereas at larger $\tilde{J}$ the amplitude of capillary waves diminishes (see Fig. 4.2a, center and bottom) and a transition occurs that delocalizes the instantaneous interface (see Fig. 4.2b, center and bottom as well as [50–55]). A sharp but delocalized instantaneous interface becomes effectively broader upon time- or ensemble-averaging over respective interface positions (see Fig. 4.2c, d). Exact results in the regime $\tilde{J} \to \infty$ for the 2D Ising strip have confirmed the interface broadening [50–53], whereas it is known that MF theories fail to account for it [54, 68]. A comprehensive theory that captures the broadening transition due to the instantaneous interface delocalization remains elusive. This example therefore motivates a deeper and more systematic analysis of interfaces and phase separation in the strong interaction limit.

4.3 Cahn-Hilliard Theory Including Pair-Correlations

Here we provide a brief summary of the derivation of the Cahn-Hilliard field theory shown in Chap. 3. For simplicity, and without much loss of generality, we limit the discussion to two-dimensional systems, i.e. $\mathbf{x} \in \mathbb{R}^2$. We start from a two-dimensional Ising model with $N_\sigma^x \times N_\sigma^y$ spins $\sigma_{ij} = \pm 1$ on a general lattice with physical dimensions L_x and L_y, respectively (see Fig. 3.1a). The Hamiltonian reads (in units of $k_{\mathrm{B}}T$)

$$\tilde{\mathcal{H}}(\boldsymbol{\sigma}) = -\tilde{J} \sum_{\langle kl,mn \rangle} \sigma_{kl}\sigma_{mn}, \tag{4.1}$$

where $\boldsymbol{\sigma}$ is the matrix containing all spin configurations, $\tilde{J} \geq 0$ is the (isotropic) interaction strength, and $\langle kl, mn \rangle$ denotes a sum over nearest neighbors. In Chap. 3 we also consider anisotropic interactions with horizontal and vertical interaction $(\tilde{J}_x, \tilde{J}_y)$.

---

[3] The instantaneous concentration refers to the concentration of down spins for a **single realization** at a **single moment in time**.

4.3.1 *From the Partition Function to the Field Theory*

We divide the lattice into $N_b^x \times N_b^y$ boxes, with sides L_x/N_b^x and L_y/N_b^y, respectively and centroids indexed by ij with $i \in \{1, \ldots, N_b^x\}$, $j \in \{1, \ldots, N_b^y\}$ (see Fig. 3.1b). We divide each spin configuration $\boldsymbol{\sigma}$ in $N_b^x \times N_b^y$ blocks encoded in the matrices $\mathbf{b}_{ij}$ with dimension $\dim \mathbf{b}_{ij} = N_\sigma^x/N_b^x \times N_\sigma^y/N_b^y$. Let $\varphi_{ij}(\mathbf{b}_{ij}) \equiv (\dim \mathbf{b}_{ij})^{-1} \sum_{\sigma_{mn} \in \mathbf{b}_{ij}} (1 - \sigma_{mn})/2$ denote the concentration of down spins in block ij (see Fig. 3.1c). For any *fixed* trial configuration of box concentrations $\hat{\varphi}_{ij} \in [0, 1]$ which we write in the matrix $\hat{\boldsymbol{\varphi}}$, the constrained partition function reads

$$Z(\hat{\boldsymbol{\varphi}}) = \sum_{\boldsymbol{\sigma}} \mathrm{e}^{-\tilde{\mathcal{H}}(\boldsymbol{\sigma})} \prod_{i=1}^{N_b^x} \prod_{j=1}^{N_b^y} \mathbb{1}_{\hat{\varphi}_{ij}}[\varphi_{ij}(\mathbf{b}_{ij})] \tag{4.2}$$

where $\mathbb{1}_x[z]$ is the indicator function of x. Generally, $Z(\hat{\boldsymbol{\varphi}})$ cannot be determined exactly. BG theory provides a variational approach to $Z(\hat{\boldsymbol{\varphi}})$ on the nearest-neighbor pair-correlation level that exactly accounts for the potential energy. Its accuracy is assessed in Sect. 3.10 for system sizes that are amenable to exact solutions. The total free energy density is evaluated as the **twofold ordered scaling limit**

$$\tilde{F}[\varphi(\mathbf{x})] \equiv \lim_{\mathrm{s}}^{\mathbf{N}_b}\left[\frac{-1}{N_b^x N_b^y} \lim_{\mathrm{s}}^{\mathbf{N}_\sigma}\left[\frac{\ln Z(\hat{\boldsymbol{\varphi}})}{(N_\sigma^x/N_b^x)(N_\sigma^y/N_b^y)}\right]\right], \tag{4.3}$$

where $\tilde{F} \equiv F/k_\mathrm{B}T$ and the scaling limit are defined in Eq. (3.1). The first limit takes the number of spins to infinity while keeping the system L_x, L_y and box $L_x/N_b^x, L_y/N_b^y$ sizes constant (see Fig. 3.1c), making $\varphi_{ij}(\mathbf{b}_{ij}) \in [0, 1]$ a continuous function in each box. The second limit takes the number of boxes to infinity while keeping L_x, L_y constant (see Fig. 3.1d), thereby making the boxes infinitesimally small and rendering $\hat{\varphi}_{ij} \to \varphi(\mathbf{x})$ a continuum field. Evaluating Eq. (4.3) analytically within the BG approximation leads to a Cahn-Hilliard functional (see Eq. (3.45))

$$\tilde{F}_{\mathrm{BG}}[\varphi(\mathbf{x})] = \frac{1}{V}\int_V \mathrm{d}\mathbf{x}\left[\tilde{\mathrm{f}}_{\mathrm{BG}}(\varphi(\mathbf{x})) + \frac{1}{2}\nabla\varphi(\mathbf{x})^T \boldsymbol{\kappa}_{\mathrm{BG}}(\varphi(\mathbf{x}))\nabla\varphi(\mathbf{x})\right], \tag{4.4}$$

with V being the system's volume, $\tilde{\mathrm{f}}_{\mathrm{BG}}(\varphi) = \tilde{\mathrm{u}}_{\mathrm{BG}}(\varphi) - \tilde{\mathrm{s}}_{\mathrm{BG}}(\varphi)$ the local free energy density given by Eq. (3.44), and $\boldsymbol{\kappa}_{\mathrm{BG}}(\varphi)$ the gradient energy coefficient given by Eq. (3.33). The MF analogs are given by Eq. (3.68) for $\tilde{\mathrm{f}}_{\mathrm{MF}}(\varphi)$, and $\lim_{\tilde{J}\to 0} \boldsymbol{\kappa}_{\mathrm{BG}}(\varphi) = \boldsymbol{\kappa}_{\mathrm{MF}} + \mathcal{O}(\tilde{J}^2)$ with $\boldsymbol{\kappa}_{\mathrm{MF}} \equiv \bar{\mathbf{z}}\tilde{J}$, where $\bar{\mathbf{z}} = \mathrm{diag}(\bar{z}_x, \bar{z}_y)$ encodes the average lattice coordination numbers in the horizontal and vertical direction. Note that $\boldsymbol{\kappa}_{\mathrm{MF}}$ is independent of φ, in agreement with regular solution theory [2]. In Fig. 4.3a we plot κ_x (i.e. the x-component of the gradient energy coefficient) for the BG and MF theory with the blue and red lines, respectively. Notably, $\kappa_{\mathrm{BG},x}(\varphi)$ displays a large *entropic* penalty of inhomogeneities at $\varphi \to 0$ and $\varphi \to 1$ that is *not* accounted for in MF theory.

The equilibrium profile of Eq. (4.4), i.e. the solution of $\delta \tilde{F}_{\mathrm{CH}}/\delta\varphi(\mathbf{x}) = 0$, solves Eq. (3.47). We now show that BG and MF theories predict starkly different behavior for moderate and strong interactions, and compare their outcomes with exact results.

4.4 Equilibrium Profile and Exact Results

We first focus on the square lattice Ising strip in Fig. 4.2c ($L_y \gg L_x$) where the magnetization varies only in the x direction, i.e. $\varphi(\mathbf{x}) = \varphi(x)$. The profile is obtained as the solution to Eq. (3.47), that we solve numerically by imposing the boundary conditions $\varphi(\pm L_x/2) = \varphi^{\pm}_{\mathrm{min}}$, where $\varphi^{-}_{\mathrm{min}} \equiv \arg\min_{0<\varphi\leq 1/2} \tilde{\mathrm{f}}(\varphi)$ and $\varphi^{+}_{\mathrm{min}} = 1 - \varphi^{-}_{\mathrm{min}}$ denote the co-existing states with compositions set by the left and right minimum of $\tilde{\mathrm{f}}(\varphi)$, respectively (see Sect. 2.9 for closed-form expressions of $\varphi^{\pm}_{\mathrm{min}}$). We fix the ensemble interface location such that $\varphi(0) = 1/2$.[4] Above the critical coupling $\tilde{J} > \tilde{J}_{\mathrm{crit}}$, where $\tilde{J}_{\mathrm{BG,crit}} \equiv \ln(\bar{z}/(\bar{z}-2))/2$ and $\tilde{J}_{\mathrm{MF,crit}} \equiv 1/\bar{z}$ (see Sects. 2.8.3 and 2.10.3), the free energy density $\tilde{\mathrm{f}}(\varphi)$ has two local minima resulting in a nonuniform $\varphi(x)$. For $\tilde{J} \leq \tilde{J}_{\mathrm{crit}}$ the profile is uniform.

4.4.1 *Comparison with Simulations*

In Fig. 3.2 we have already seen that the BG interface steepness scales non-monotonically w.r.t. $\tilde{J}$. To better comprehend the origin of this unexpected scaling, we compare our results with MC simulations of the two-dimensional Ising model (for simulation details see Appendix 1). In Fig. 4.2c, d we find that the BG concentration profiles are very close to the simulation results. Not only does the BG approximation correctly capture the non-monotonic trend of the interface steepness (see Fig. 4.2d), it also captures remarkably well the full concentration profile (see Fig. 4.2c). By comparing with Fig. 4.2b we observe a strong correlation between interface broadening and interface delocalization. This is further analyzed in Fig. 4.3.

4.4.2 *Comparison with Exact Results*

4.4.2.1 Scaling of Interface Steepness with the Horizontal Lattice Side L_x

First, we inspect in Fig. 4.3b the interface steepness $\varphi'(0)$ *for various horizontal lattice sides* L_x. In stark contrast to MF theory predicting a steepening interface

[4] Fixing the ensemble averaged interface position is *not* equal to fixing the position of instantaneous profiles. Thus, the interface location along individual trajectories may still fluctuate.

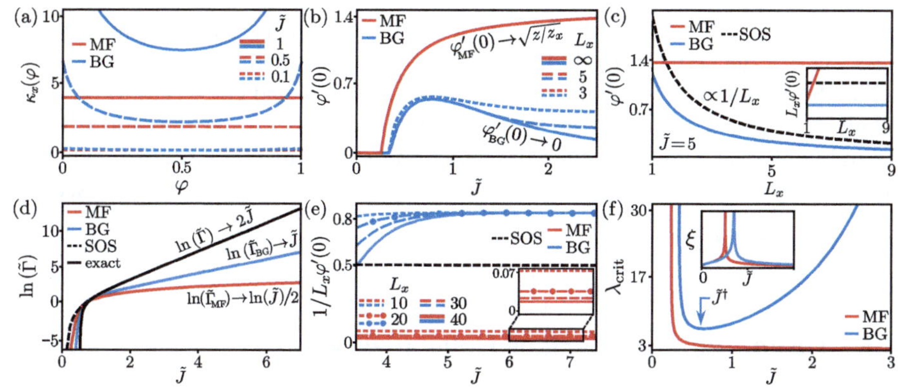

Fig. 4.3 In all panels we consider an Ising strip with $\bar{z} = 4$ and $\bar{z}_x = 2$. Red, blue, and black solid/dashed lines always correspond to MF, BG, and exact/SOS results, respectively. **a** x-component of the gradient energy coefficient, κ_x, in Eq. (3.33) as a function of φ for $\tilde{J} \in \{0.1, 0.5, 1\}$. **b, c** Interface steepness $\varphi'(0) = \partial_x\varphi|_{x=0}$ of the equilibrium concentration profile as a function of $\tilde{J}$ for fixed $L_x \in \{3, 5, \infty\}$ (**b**) and as a function of L_x for fixed $\tilde{J} = 5$ (**c**). Inset of **c** rescaled steepness $L_x\varphi'(0)$. **d** Interface stiffness $\tilde{\Gamma}$ defined in Eq. (4.5) as a function of $\tilde{J}$ on a logarithmic scale. **e** $1/L_x\varphi'(0)$ as a function of $\tilde{J}$ for fixed $L_x \in \{10, 20, 30, 40\}$. Blue lines converge to the value $\delta_{\text{BG}} \approx 0.835$. Inset: blow-up of the MF result. **f** Critical stability wavelength $\lambda_{\text{crit}} = 2\pi[-\kappa_x(1/2)/\tilde{f}''(1/2)]^{1/2}$ as a function of $\tilde{J}$; the blue arrow indicates $\tilde{J}^\dagger(1/2)$ in Eq. (3.61), where $\lambda_{\text{crit}}^{\text{BG}}$ attains a minimum. Inset: bulk correlation length $\xi \equiv [\kappa_x(\varphi_{\min}^{\pm})/\tilde{f}''(\varphi_{\min}^{\pm})]^{1/2}$

independent of lattice size, BG profiles are non-monotonic in $\tilde{J}$ beyond a sufficient L_x due to interface delocalization (Fig. 3.2 shows similar results). To verify that this is *no artifact*, we compare our results with the **SOS** model for the square lattice Ising strip ($\bar{z} = 4$), which becomes exact in the limit $\tilde{J} \to \infty$ and is known to include interface delocalization [50–54]. The SOS model yields $\lim_{\tilde{J}\to\infty} \varphi(x) = 1/2 + x/L_x + \sin(2\pi x/L_x)/2\pi$ [51, 52, 54], hence $\lim_{\tilde{J}\to\infty} \varphi'(0) = 2/L_x$. In Fig. 4.3c we show the interface steepness as a function of L_x for fixed $\tilde{J}$, and find that the SOS and BG results display the same scaling (see Fig. 4.3c inset), whereas the MF result is in fact independent of L_x.

4.4.2.2 Interface Stiffness

Further verification is given by the **interface stiffness**, which is the free energy difference between the non-uniform equilibrium profile $\varphi(x)$ and a uniform profile $\varphi_{\min}^{\pm}$, and reads (see Eq. (2.15) in [2])

$$\tilde{\Gamma} = 2 \int_{\varphi_{\min}^{-}}^{\varphi_{\min}^{+}} [\kappa_x(\varphi)(\tilde{f}(\varphi) - \tilde{f}(\varphi_{\min}^{\pm}))]^{1/2} d\varphi, \tag{4.5}$$

which is depicted in Fig. 4.3d. Note that surface tension σ is related to surface stiffness via $\sigma = \text{arcsinh}\,(\tilde{\Gamma})$ [51]. The exact result is $\tilde{\Gamma} = \sinh(2\tilde{J} + \ln\tanh\tilde{J})$ [51], while the SOS model yields $\tilde{\Gamma}_{\text{SOS}} = \cosh(2\tilde{J}) - 1$ [51] and converges to the exact result for large $\tilde{J}$, i.e. $\lim_{\tilde{J}\to\infty} \ln(\tilde{\Gamma}_{\text{SOS}}) \simeq 2\tilde{J}$. Notably, the BG result is not only considerably more accurate than the MF prediction, but also displays a correct exponential scaling, $\lim_{\tilde{J}\to\infty} \ln(\tilde{\Gamma}_{\text{BG}}) \simeq \tilde{J}$, in stark contrast to the square-root MF scaling, $\lim_{\tilde{J}\to\infty} \ln(\tilde{\Gamma}_{\text{MF}}) \simeq \ln(\sqrt{\tilde{J}})$.

4.5 Disentangling Interface Delocalization

By exploiting the mapping of instantaneous interface positions onto a **Brownian excurison problem** (see Appendix 2), we can disentangle interface delocalization from the **instantaneous interface width**, δ, in the large $\tilde{J}$ limit where the instantaneous interface positions become asymptotically uniformly distributed, i.e. $\lim_{\tilde{J}\to\infty} p_{\text{int}}(x; \tilde{J}) = L_x^{-1}\mathbb{1}_{|x|<L_x/2}$ with $\mathbb{1}_{|x|<L_x/2}$ equal to 1 when $|x| < L_x/2$ and 0 otherwise (see derivation in Appendix 2.2). For $\tilde{J} \gg 1$ each instantaneous profile $\varphi_{\text{j}}(x)$ corresponds to some continuous function $f(x/\delta + b_j) : \mathbb{R} \to [0, 1]$ obeying $\lim_{x\to\pm\infty} f(x) = (1 \pm 1)/2$, where $\delta > 0$ and b_j describe the width and position of the instantaneous interface j. The ensemble-averaged profile is then given by the convolution of $f(x)$ with the probability density to have a certain shift b, i.e.,

$$\lim_{\tilde{J}\to\infty} \varphi(x) = L_x^{-1} \int f(x/\delta + b)\mathbb{1}_{|b|<L_x/2} db.$$

We can now straightforwardly compute the interface steepness and find $\lim_{\tilde{J}\to\infty} \varphi'(0) = \delta^{-1}(f(L_x/2) - f(-L_x/2))/L_x$. Finally, taking the large-L_x, limit we obtain

$$\lim_{L_x\to\infty} \lim_{\tilde{J}\to\infty} 1/L_x\varphi'(0) = \delta, \tag{4.6}$$

and have thereby disentangled interface delocalization from the instantaneous interface width δ. This result holds for any continuous integrable $f(x)$ obeying $\lim_{x\to\pm\infty} f(x) = (1 \pm 1)/2$. For the Ising strip this yields $\delta_{\text{SOS}} = 0.5$ within the SOS model, and $\delta_{\text{BG}} \approx 0.835$ with the BG approximation (see Fig. 4.3e). Hence, we conclude:

> While interface delocalization causes the ensemble averaged steepness to vanish in the large coupling limit, individual realizations maintain a nonzero interface steepness with uniformly distributed instantaneous interface positions.

Importantly, MF theory does *not* account for delocalization-induced interface broadening and therefore predicts $\delta_{\text{MF}} \to 0$ (see inset of Fig. 4.3e).

4.6 Spinodal Decomposition

Having established the physical consistency of the BG field theory given by Eq. (3.45), we now address phase separation, and determine the length scales on which inhomogeneities are stable by performing a linear stability analysis on the total free energy density around the uniform concentration profile, $\varphi(x) = \varphi_0 + a\sin(qx)$ with $|a| \ll \min(\varphi_0, 1-\varphi_0)$ (the symmetry of the problem imposes odd inhomogeneities). Stable perturbations lower the total free energy density, $\Delta\tilde{F}_{\mathrm{BG}} \equiv \tilde{F}_{\mathrm{BG}}[\varphi(x)] - \tilde{F}_{\mathrm{BG}}[\varphi_0] \leq 0$, yielding an upper bound on stable wavevectors $q \leq q_{\mathrm{crit}}^{\mathrm{BG}} \equiv [-\tilde{\mathrm{f}}''_{\mathrm{BG}}(\varphi_0)/\kappa_{\mathrm{BG},x}(\varphi_0)]^{1/2}$ (see derivation in Sect. 3.8), where $\tilde{\mathrm{f}}''_{\mathrm{BG}}(\varphi) = \mathrm{d}^2\tilde{\mathrm{f}}_{\mathrm{BG}}(\varphi)/\mathrm{d}\varphi^2$ is the curvature of the free energy barrier given by Eq. (2.43). The critical wavevector $q_{\mathrm{crit}}^{\mathrm{BG}}$ translates into a critical wavelength $\lambda_{\mathrm{crit}}^{\mathrm{BG}} = 2\pi/q_{\mathrm{crit}}^{\mathrm{BG}}$ above which perturbations are stable. Fig. 4.3f depicts λ_{crit} as a function of $\tilde{J}$ for a square lattice with $\varphi_0 = 1/2$ (for general φ_0 see Fig. 3.3c, d). Similar to Fig. 4.3b, λ_{crit} displays a non-monotonic trend in the BG theory (blue lines) that is contrasted by a monotonic attenuation in the MF theory (red lines). The non-monotonicity of $\lambda_{\mathrm{crit}}^{\mathrm{BG}}$ is explained by inspecting how the curvature of the barrier depends on $\tilde{J}$. In particular, the BG curvature converges, $\lim_{\tilde{J}\to\infty} \tilde{\mathrm{f}}''_{\mathrm{BG}}(1/2) = 2(2-\bar{z})$ (see Fig. 2.6c, blue line), whereas the free energy penalty of inhomogeneities $\kappa_{\mathrm{BG},x}$ increases exponentially, eventually increasing $\lambda_{\mathrm{crit}}^{\mathrm{BG}}$. MF theory overestimates the curvature of the barrier (see Fig. 2.6c, red line), and underestimates the free energy penalty of inhomogeneities, leading to a decreasing $\lambda_{\mathrm{crit}}^{\mathrm{MF}}$. The **bulk correlation length** $\xi \equiv [\kappa_x(\varphi_{\mathrm{min}}^{\pm})/\tilde{\mathrm{f}}''(\varphi_{\mathrm{min}}^{\pm})]^{1/2}$ [42] displays qualitatively the same behavior in both theories (see Fig. 4.3f, inset), since the MF free energy density is relatively accurate near local minima $\varphi_{\mathrm{min}}^{\pm}$ but inaccurate near the barrier.

4.7 Nucleation

We next investigate, in Fig. 4.4, how interface broadening affects nucleation by determining minimal free energy paths (the reaction coordinate and numerical method are described in Appendix 3). The inset in Fig. 4.4a suggests that critical nuclei become less dense and wider as $\tilde{J}$ becomes larger. Indeed, we find that correlations captured by BG theory lead to larger critical nuclei (Fig. 4.4b), shallower interfaces (Fig. 4.4c), and that the increasing trend with $\tilde{J}$ is only captured by BG theory, which is reminiscent of the results shown in Fig. 4.3. Most importantly, BG theory predicts that the nucleation barrier ΔE is approximately four times larger than predicted by MF (Fig. 4.4a), implying a strong reduction of nucleation rates [69–72].

To understand why interface delocalization affects nucleation, we note that shifting the interface position corresponds to a growing/shrinking nucleus, which consequently alters the free energy. Instantaneous interfaces are still affected by interface translation and capillary-wave fluctuations. However, in contrast to the strip, distinct instantaneous interface configurations are *not* iso-energetic. The weighting by the

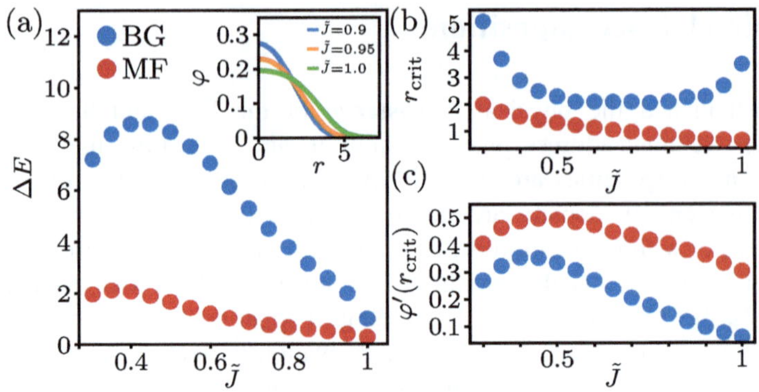

Fig. 4.4 Numerical simulations of critical nuclei of the radially symmetric Cahn-Hilliard equation with the BG (blue) and MF (red) free energy for a hexagonal coordination ($\bar{z} = 6$, $\bar{z}_x = 4$). **a** Free energy difference ΔE between the critical nucleus and the homogeneous state as a function of the interaction strength $\tilde{J}$. The inset shows critical profiles $\varphi(r)$ for three values of $\tilde{J}$. **b** Radius r_{crit} and **c** interface steepness $\varphi'(r_{\text{crit}})$ of the critical nucleus as a function of $\tilde{J}$

respective free energy of the configuration ultimately gives rise to broadening and thus larger critical nuclei and higher nucleation barriers.

4.8 Concluding Remarks

In Chap. 3 we derived a Cahn-Hilliard field theory that accounts for nearest-neighbor pair-correlations, by directly computing the thermodynamic limit of a spatially inhomogeneous two-dimensional Ising model within the BG approximation. Here we have shown that strong interactions give rise to (i) a **delocalization-induced interface broadening** confirmed by MC simulations and exact results for the two-dimensional Ising model, (ii) a **strong reduction of nucleation kinetics** due to an amplification of the free energy barrier to nucleation, and (iii) a **non-monotonic dependence of critical nucleus size** on interaction strength. These effects are the result of an entropy-driven interplay between capillary-wave and interface-position fluctuations at sufficiently strong coupling, and pair correlations are required to correctly account for them. Pair correlations enforce a thermodynamically optimal configuration of defects, and are thus an essential determinant of interfaces and condensates in the strong interaction limit that so far have been overlooked. Our results allow for generalizations to three dimensions and more than two constituents (i.e. ternary mixtures), which will be addressed in future work.

Appendices

The Appendices are organized in the order they appear in this chapter. First, we present in Appendix 1, a detailed description of MC simulations that are shown in Fig. 4.2. In Appendix 2 we proceed with a derivation of the probability density of instantaneous interface positions, based on the mapping onto the Brownian bridge problem (the results are briefly shown in Fig. 4.2b). Finally, in Appendix 3 we present details on the numerical simulations of nucleation by means of the radially symmetric Cahn-Hilliard equation which are shown in Fig. 4.4. Any information and detailed calculations related to the Cahn-Hilliard free energy functional within the BG and MF approximation are provided in Chap. 3.

Appendix 1: Monte-Carlo Simulations of the Ising Model

Here we provide details on the MC simulations we performed to determine the ensemble averaged concentration profile and histograms of instantaneous interface locations displayed in Fig. 4.2.

Appendix 1.1: Lattice Setup and Initial Configuration

We performed MC simulations of the nearest-neighbor interacting ferromagnetic Ising model on the square lattice with size ($N_\sigma^x = 40$) $\times$ ($N_\sigma^y \in \{80, 90, 100, 110, 120, 130\}$) with single spin-flip dynamics in the bulk and two-spin-exchange dynamics at the boundary columns located at $i = \pm N_\sigma^x/2$. We considered various values of N_σ^y to benchmark our simulations against known theoretical predictions of the scaling behavior (see Appendix 1.5). We imposed periodic boundary conditions in the vertical direction (i.e. along the columns), and free boundary conditions in the horizontal direction (i.e. along the rows), whereby we constrained the total magnetization on the left/right boundary (see below). Let $N_i^\downarrow$ with $i \in \{-N_\sigma^x/2, ..., N_\sigma^x/2\}$ denote the number of down spins in column i. To induce a non-uniform concentration profile, and in anticipation of known exact results for the bulk concentration values [73], we fixed the number of down spins at the boundaries to be

$$N^\downarrow_{\pm N_\sigma^x/2} = \frac{N_\sigma^y}{2}\left(1 \pm \mathrm{Re}([1 - \sinh^{-4}(2\tilde{J})]^{1/8})\right), \tag{4.7}$$

where $\tilde{J}$ is the coupling strength and $\mathrm{Re}(\cdot)$ denotes the real part. Spins located at the boundaries can exchange *only* within the same column, and therefore the total number of up/down spins at the boundaries is conserved throughout the simulation. Spins in the bulk are initially prepared in a high-coupling configuration (i.e. aligned) with a vertical interface placed at some random horizontal location in the lattice. Starting from a high-coupling configuration has the advantage that the simulations do not get stuck in frozen sub-optimal states where multiple interfaces are created [74, 75].

Appendix 1.2: Acceptance Rate

For single spin-flip dynamics let $\{\sigma_j\}'_i$ denote the spin configuration obtained by flipping spin i while keeping the configuration of all other spins fixed, i.e., $\{\sigma_j\}'_i \equiv (-\sigma_i, \{\sigma_{j\neq i}\})$. Moreover, let $p_i(\{\sigma_j\})$ denote the acceptance rate from $\{\sigma_j\}$ to $\{\sigma_j\}'_i$ and $\Delta\mathcal{H}_i(\{\sigma_j\}) \equiv \mathcal{H}(\{\sigma_j\}'_i) - \mathcal{H}(\{\sigma_j\})$ the energy difference associated with the transition. Using the **Metropolis algorithm** the acceptance rate for the single spin-flip takes the form [76]

$$p_i(\{\sigma_j\}) = \min(1, \mathrm{e}^{-\Delta\mathcal{H}_i(\{\sigma_j\})}). \tag{4.8}$$

For two-spin-exchange dynamics let $\{\sigma_j\}'_{ik}$ denote the spin configuration upon interchanging the spins σ_i and σ_k while keeping the configuration of all other spins fixed, i.e., $\{\sigma_j\}'_{ik} \equiv (\sigma_i \leftrightarrow \sigma_k, \{\sigma_{j\neq(i,k)}\})$. We denote with $p_{ik}(\{\sigma_j\})$ the acceptance rate from $\{\sigma_j\}$ to $\{\sigma_j\}'_{ik}$ and $\Delta\mathcal{H}_{ik}(\{\sigma_j\}) \equiv \mathcal{H}(\{\sigma_j\}'_{ik}) - \mathcal{H}(\{\sigma_j\})$ denotes the energy difference associated with the transition. Using the Metropolis algorithm the two-spin-exchange acceptance rate reads

$$p_{ik}(\{\sigma_j\}) = \min(1, \mathrm{e}^{-\Delta\mathcal{H}_{ik}(\{\sigma_j\})}). \tag{4.9}$$

Appendix 1.3: Simulation Parameters

For each value of the coupling strength $\tilde{J}$ and vertical length $N_\sigma^y \in \{80, 90, 100, 110, 120, 130\}$ we performed $N_{\mathrm{MC}} = 10^5$ MC simulations, where each individual run contained 5×10^8 MC steps. At each 1.9×10^7th MC step we took a snapshot of the configuration and stored the total energy, resulting in 26 (including the initial configuration) snapshots for each simulation run.

Appendix 1.4: Equilibration Test: Energy Fluctuations Per Spin

To assess whether the MC simulations reached equilibrium, we analyzed the energy fluctuations per spin and their corresponding ensemble average. In Fig. 4.5 we display the energy fluctuations per spin for a subset of 10^4 simulations as a function of the MC

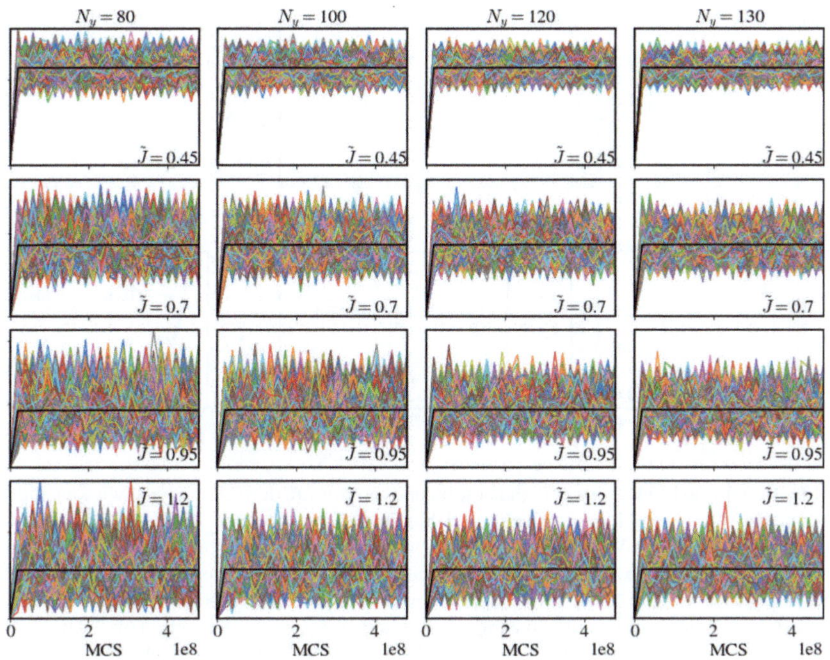

Fig. 4.5 **Equilibration test**: energy fluctuations per spin as a function of consecutively stored MC configurations. In each plot we display the energy fluctuations per spin $(\mathrm{E} - \mathrm{E}_0)/N$, where E_0 is the ground state energy conditioned on anti-symmetric boundary conditions, and $N = N_\sigma^x N_\sigma^y$ with $N_\sigma^x = 40$ for a subset of 10^4 MC simulations (colored lines). The black solid line indicates the ensemble average energy fluctuation per spin. Plots in the same column have equal $N_\sigma^y \in \{80, 100, 120, 130\}$, and plots in the same row have equal $\tilde{J} \in \{0.45, 0.6, 0.95, 1.2\}$

steps for various $\tilde{J} \in \{0.45, 0.7, 0.95, 1.2\}$ and $N_\sigma^y \in \{80, 100, 120, 130\}$. In each plot we observe that immediately after the initial snapshot the energy is fluctuating around an average steady state denoted with the black solid line, providing a first indication that the simulations have reached equilibrium (already at the first stored configuration). Note that in each plot all energies are initially increasing from zero since we subtract the ground state energy *and* we initialize the system in a high-coupling configuration which is identical to the ground state.

Appendix 1.5: Benchmark Test: Interface Width and Roughening

To benchmark the performance of our MC simulations, we computed the interface width $w^2(N_\sigma^y, \tilde{J})$ and compared our results with known theoretical results reported in [43, 77]. The results from [77] predict $w^2(N_\sigma^y, \tilde{J}) \propto N_\sigma^y / \sinh(\sigma)$

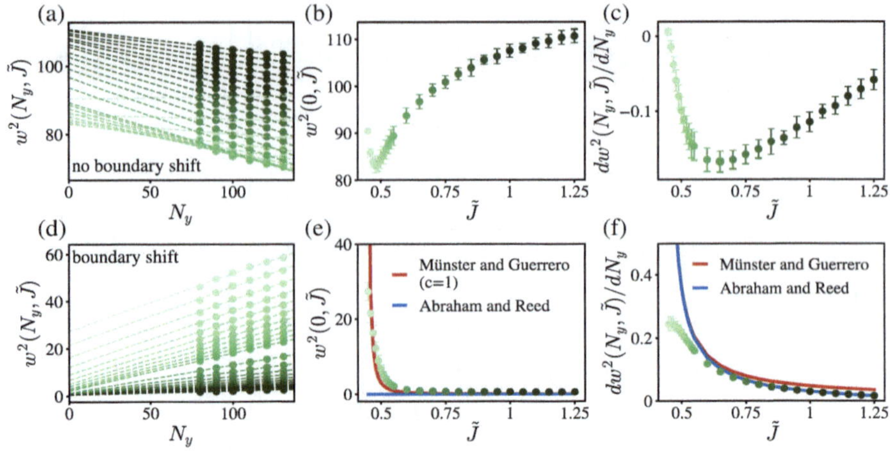

Fig. 4.6 **Benchmark test**: results in the bottom and top row are derived with and without applying the boundary-shift method, respectively. **a–d** Scaling of the interface width (**a**) $w^2(N^y_\sigma, \tilde{J})$ (no boundary shift) and **d** $\hat{w}^2(N^y_\sigma, \tilde{J})$ (boundary shift) w.r.t. the vertical number of spins N^y_σ. Each point is obtained by averaging over 2.5×10^6 equilibrated configurations. Dashed lines are obtained by weighted linear regression. Colors from light green to dark green correspond to increasing coupling strength $\tilde{J}$. **b–e** Intersection point of the interfacial width at $N^y_\sigma = 0$ as a function of $\tilde{J}$. The standard deviation of each point is estimated with the Jackknife method. In **e** the red and blue lines are the theoretical predictions for the intersection point given in [43, 77], respectively. **c–f** Slope of the interface width w.r.t. N^y_σ as a function of $\tilde{J}$. The standard deviation of each point is estimated with the Jackknife method. In **f** the red and blue lines are the theoretical predictions for the slope given in [43, 77], respectively

with $\sigma = 2\tilde{J} + \ln\tanh\tilde{J}$. Analogously, the results from [43] predict $w^2(N^y_\sigma, \tilde{J}) = N^y_\sigma/12\sigma - c/2\pi\sigma^2$ with $c \approx 1$. Below we explain in detail how we determined the interface width and how it compares to the theoretical predictions. The resulting outcomes are shown in Fig. 4.6 and the comparison with the theoretical results are shown in Fig. 4.6e–f.

Ensemble Averaged Concentration Profile and the Boundary-Shift Method

To compare our results with [43, 77], we need to apply the so-called **boundary shift method** [78], where we shift the interface position of each instantaneous concentration profile to the center of the lattice. As a scientific exercise we also consider the resulting outcomes without applying the boundary shift method, for which the results are depicted in the top row of Fig. 4.6. Let $\hat{\varphi}_{i,k}$ be the equilibrated and boundary shifted concentration of down spins in column $i \in \{-N^\sigma_x/2, \ldots, N^\sigma_x/2\}$ of the kth MC simulation run. We define the ensemble average boundary-shifted concentration profile as

$$\langle\hat{\varphi}_i\rangle = \frac{1}{N_{\mathrm{MC}}}\sum_{k=1}^{N_{\mathrm{MC}}}\hat{\varphi}_{i,k}. \tag{4.10}$$

From Eq. (4.10) we can approximate the mean interface width using the central difference method as follows: Let $\delta_2[\langle\varphi_i\rangle] \equiv \langle\varphi_{i+1}\rangle - \langle\varphi_{i-1}\rangle$ then

$$\hat{w}^2(N_\sigma^y, \tilde{J}) = \frac{\sum_{i=-N_\sigma^x/2+1}^{N_\sigma^x/2-1} i^2\delta_2[\langle\varphi_i\rangle]}{\sum_{i=-N_\sigma^x/2+1}^{N_\sigma^x/2-1} \delta_2[\langle\varphi_i\rangle]} - \left(\frac{\sum_{i=-N_\sigma^x/2+1}^{N_\sigma^x/2-1} i\delta_2[\langle\varphi_i\rangle]}{\sum_{i=-N_\sigma^x/2+1}^{N_\sigma^x/2-1} \delta_2[\langle\varphi_i\rangle]}\right)^2. \tag{4.11}$$

A similar definition holds for the interface width without applying the boundary-shift method, which we denote as $w^2(N_\sigma^y, \tilde{J})$. In Fig. 4.6 we plot $w^2(N_\sigma^y, \tilde{J})$ and $\hat{w}^2(N_\sigma^y, \tilde{J})$ with the green dots as a function of N_σ^y. Both results show a clear linear trend with N_σ^y, providing a first validation of the MC simulations. To obtain the variance of $w^2(N_\sigma^y, \tilde{J})$—which we use in the next section for weighted linear regression—we used the Jackknife method which is explained below.

Interface Width and Weighted Linear Regression

To compare our results with those reported in [43, 77], we need to extract the interception point $\hat{w}^2(\tilde{J}, 0)$ and slope $d\hat{w}^2(\tilde{J}, N_\sigma^y)/dN_\sigma^y$. Both quantities are obtained through weighted linear regression in combination with the Jackknife method. First we determine $\hat{w}^2(0, \tilde{J})$ and $d\hat{w}^2(N_\sigma^y, \tilde{J})/dN_\sigma^y$ for fixed $\tilde{J}$ while removing one point from the data pool, which gives

$$\left\{\hat{w}_j^2(0, \tilde{J}), \frac{d\hat{w}_j^2(N_\sigma^y, \tilde{J})}{dN_\sigma^y}\right\} = \min_{(\alpha,\beta)} \sum_{\substack{N_\sigma^y=\{80,\ldots,130\} \\ N_\sigma^y \neq 70+10\times j}} \frac{(\alpha + \beta N_\sigma^y - \hat{w}^2(N_\sigma^y, \tilde{J}))^2}{\mathrm{var}(\hat{w}^2(N_\sigma^y, \tilde{J}))}, \tag{4.12}$$

where $j = \{1, ..., 6\}$. A similar definition holds for the intersection point and slope without applying the boundary shift method, which we denote as $w_j^2(0, \tilde{J})$ and $dw_j^2(N_\sigma^y, \tilde{J})/dN_\sigma^y$, respectively. Finally, the Jackknife ensemble averages and variances are given by

$$\hat{w}^2(0, \tilde{J}) = \frac{1}{6}\sum_{j=1}^{6} \hat{w}_j^2(0, \tilde{J}), \; \mathrm{var}(\hat{w}^2(0, \tilde{J})) = \frac{5}{6}\sum_{j=1}^{6}(\hat{w}_j^2(0, \tilde{J}) - \hat{w}^2(0, \tilde{J}))^2,$$

$$\frac{d\hat{w}^2(N_\sigma^y, \tilde{J})}{dN_\sigma^y} = \frac{1}{6}\sum_{j=1}^{6} \frac{d\hat{w}_j^2(N_\sigma^y, \tilde{J})}{dN_\sigma^y},$$

$$\mathrm{var}\left(\frac{d\hat{w}^2(N_\sigma^y, \tilde{J})}{dN_\sigma^y}\right) = \frac{5}{6}\sum_{j=1}^{6}\left(\frac{d[\hat{w}_j^2(N_\sigma^y, \tilde{J}) - \hat{w}^2(N_\sigma^y, \tilde{J})]}{dN_\sigma^y}\right)^2.$$

In Fig. 4.6e–f we plot $\hat{w}^2(0, \tilde{J})$ and $d\hat{w}^2(N_\sigma^y, \tilde{J})/dN_\sigma^y$ together with the standard deviation as a function of $\tilde{J}$. The theoretical results given by [43, 77] are shown with the red and blue lines, respectively. For $\tilde{J} \geq 0.6$ we find a very good agreement between MC simulations and theoretical predictions. Notably, for the slope in Fig. 4.6f we find a remarkable agreement with the results of [77]. For $\tilde{J} < 0.6$ we approach the *exact* critical coupling $\tilde{J}_{\text{crit}} \approx 0.441$, where the MC results agree less well with theoretical predictions due to finite-size effects. This is expected since the correlation length diverges around the critical coupling.

Appendix 2: Statistics of Instantaneous Interface Position

Neglecting overhangs, one can map the statistics of instantaneous interfaces onto a one-dimensional confined Brownian bridge problem [55]. The idea is to treat the respective bulk phases as "pure" (i.e. homogeneous) and the interface (i.e. domain wall) as a random walk, which in the continuum limit becomes a Brownian motion where the vertical coordinate y plays the role of time and the diffusion coefficient is proportional to $1/4\tilde{\Gamma}$, where $\tilde{\Gamma} = \sinh(2\tilde{J} + \ln\tanh\tilde{J})$ is the exact interface stiffness for the two-dimensional Ising model [77, 79]. Periodic boundary conditions in the y-direction render the Brownian trajectories Brownian bridges. Here we derive the probability density of instantaneous interface positions, based on the analogy with Brownian bridges, used for Fig. 4.2b. Furthermore, we show the convergence to a uniform distribution in the limit $\tilde{J} \to \infty$, which we used for Sect. 4.5 to disentangle interface delocalization from the instantaneous interface width.

Appendix 2.1: Derivation

Let us assume that the instantaneous interface in the two-dimensional Ising strip always separates two homogeneous thermodynamically co-existing phases and that the domain wall behaves like a Gaussian polymer confined between hard walls located at $x = 0$ and $x = L_x$ with a height $L_y >> L_x$. We parameterize the domain wall as a Brownian motion $\{x_y^{\mathrm{i}}\}_{0 \le y \le L_y}$ where y plays the role of time (or contour length in the polymer context). Then, the Green's function of the interface with diffusion coefficient D follows the **Edwards equation** with absorbing boundary conditions at the walls

$$
\begin{aligned}
\partial_y G(x, y|x_0) &= D\partial_x^2 G(x, y|x_0), \\
G(x, 0|x_0) &= \delta(x - x_0), \\
G(0, y|x_0) &= G(L_x, y|x_0) = 0, \ \forall\, y \in [0, L_y].
\end{aligned}
\tag{4.13}
$$

The general solution to Eq. (4.13) is

$$G(x, y|x_0) = \frac{2}{L_x} \sum_{k=1}^{\infty} \sin(k\pi x/L_x) \sin(k\pi x_0/L_x) e^{-\pi^2 k^2 Dy/L_x^2}. \tag{4.14}$$

Particularly interesting is the mean squared displacement (MSD) which is given by

$$\begin{aligned} \langle (x-x_0)^2 \rangle &= \frac{1}{L_x} \int_0^{L_x} dx \int_0^{L_x} dx_0 (x-x_0)^2 G(x, y|x_0) \\ &= \frac{4L_x^2}{\pi^4} \sum_{k=1}^{\infty} \frac{(-1)^{k+1}(k^2\pi^2-4)-4}{k^4} e^{-\pi^2 k^2 Dy/L_x^2}. \end{aligned}$$

For $y << D/L_x^2$ we can truncate the exponential inside the sum up to first order in y and obtain $\langle (x - x_0)^2 \rangle \simeq 2Dy$, as expected for a freely diffusing one-dimensional particle. Now we recall the exact results of Abraham [77] and Fisher [79] who found that for the two-dimensional Ising strip the interface width should scale as[5] $\langle (x - x_0)^2 \rangle \propto y/2\tilde{\Gamma}$ where the proportionality factor includes some lattice length scale and $\tilde{\Gamma} = \sinh(2\tilde{J} + \ln \tanh \tilde{J})$ is the surface stiffness as defined in Eq. (4.5). The outcome of [77, 79] allows us to relate the diffusion coefficient D to the surface stiffness

$$D \propto \frac{1}{4 \sinh(2\tilde{J} + \ln \tanh \tilde{J})}. \tag{4.15}$$

Under periodic boundary conditions in the y-direction the interfaces have an equal position at $y = 0$ and $y = L_y$ (also known as Brownian bridges). In this case the propagator is simply given by $G(x, L_y|x)$. We can now calculate the probability density to have an interface located at position x, which upon normalization is given by

$$p_{\mathrm{int}}(x; \tilde{J}) = \frac{G(x, L_y|x)}{\int_0^{L_x} G(x, L_y|x) dx} = \frac{1}{L_x} \frac{\vartheta_3(0, e^{-\alpha_{\tilde{J}}}) - \vartheta_3(\pi x/L_x, e^{-\alpha_{\tilde{J}}})}{\vartheta_3(0, e^{-\alpha_{\tilde{J}}}) - 1}, \tag{4.16}$$

where $\alpha_{\tilde{J}} \equiv \pi^2 D L_y / L_x^2$ and $\vartheta_3(a, x)$ is Jacobi's elliptic theta of the third kind. The second equality can be obtained directly from Eq. (4.14) with the definition of Jacobi's elliptic theta of the third kind. Plugging Eq. (4.15) for the diffusion coefficient into Eq. (4.16), we can compute the black dashed lines in Fig. 4.2b. As long as $\tilde{J} \ll 1$ we have $\tilde{\Gamma} \ll 1$ and thus $\alpha_{\tilde{J}} \gg 1$. Accordingly, Eq. (4.16) predicts instantaneous interfaces to be localized with a probability density $p_{\mathrm{int}}(x; \tilde{J}) \propto \sin(\pi x/L_x)^2$ (see

[5] The vertical dimension in [77] is defined from $y = [-L_y, L_y]$, hence in our terminology there is an additional factor of $1/2$ in the MSD.

Fig. 4.2b, top panel). Conversely, for sufficiently large $\tilde{J}$ we find $\tilde{\Gamma} \gg 1$ and hence $\alpha_{\tilde{J}} \ll 1$, and the interface positions become delocalized (see Fig. 4.2b, center and bottom panels).

Appendix 2.2: Convergence to the Uniform Distribution

Next we prove that Eq. (4.16) converges to the uniform distribution in the limit $\tilde{J} \to \infty$. We first introduce $q \equiv \mathrm{e}^{-\alpha_{\tilde{J}}}$ and rewrite

$$\vartheta_3(\pi x/L_x, \mathrm{e}^{-\alpha_{\tilde{J}}}) = 1 + 2\sum_{n=1}^{\infty} \mathrm{e}^{-n^2\alpha_{\tilde{J}}} \cos(2\pi n x/L_x) = \sum_{n=-\infty}^{\infty} q^{n^2} \mathrm{e}^{2\pi n x \mathrm{i}/L_x}. \quad (4.17)$$

Since $\alpha_{\tilde{J}} \geq 0$ and $\lim_{\tilde{J}\to\infty} \alpha_{\tilde{J}} = 0$ (uniformly), we find that $\lim_{\tilde{J}\to\infty} \mathrm{e}^{-\alpha_{\tilde{J}}}$ is equivalent to $\lim_{q\uparrow 1} q$ in Eq. (4.17). We now use the asymptotic result for $q \uparrow 1$ [80]

$$\lim_{q\uparrow 1} \sum_{n=-\infty}^{\infty} q^{n^2} \left(\mathrm{e}^{\mathrm{i}2\pi x/L_x}\right)^n \simeq \sqrt{\frac{\pi}{-\ln q}} \exp\left(\frac{\pi^2 x^2}{L_x^2 \ln q}\right), \quad (4.18)$$

where $\simeq$ stands for asymptotic equality, i.e. $A \simeq B$ stands for $A/B \to 1$. Note that Eq. (4.18)$\gg 1$ since $0 < x < L_x$. Let us now rewrite Eq. (4.16) as

$$L_x p_{\mathrm{int}}(x; \tilde{J}) = \frac{\vartheta_3(0, \mathrm{e}^{-\alpha_{\tilde{J}}}) - 1 - \vartheta_3(\pi x/L_x, \mathrm{e}^{-\alpha_{\tilde{J}}}) + 1}{\vartheta_3(0, \mathrm{e}^{-\alpha_{\tilde{J}}}) - 1} = 1 - \frac{\vartheta_3(\pi x/L_x, \mathrm{e}^{-\alpha_{\tilde{J}}}) - 1}{\vartheta_3(0, \mathrm{e}^{-\alpha_{\tilde{J}}}) - 1} \quad (4.19)$$

We can now evaluate the limit of Eq. (4.19) using Eq. (4.18), and find

$$\lim_{\tilde{J}\to\infty} L_x p_{\mathrm{int}}(x; \tilde{J}) \simeq 1 - \lim_{q\uparrow 1} \exp\left(\frac{\pi^2 x^2}{L_x^2 \ln q}\right) = 1 - \lim_{\alpha_{\tilde{J}}\downarrow 0} \exp\left(-\frac{\pi^2 x^2}{L_x^2 \alpha_{\tilde{J}}}\right) \to 1 \quad (4.20)$$

for $0 < x < L_x$, while we have $p_{\mathrm{int}}(0; \tilde{J}) = p_{\mathrm{int}}(L_x; \tilde{J}) = 0$, $\forall \tilde{J}$. Applying a translational shift $x \to x - L_x/2$ yields the resulting uniform distribution mentioned Sect. 4.5.

Notably, when $\tilde{J} \to \infty$ a Casimir effect appears in addition (see e.g. [81, 82]) that is *not* captured in Eq. (4.16), i.e. the entropy due to *bulk* fluctuations is enhanced near the boundaries giving rise to "peaks" (see Fig. 4.2b, bottom panel).

Appendix 3: Simulating the Radially Symmetric Cahn-Hilliard Equation

We study nucleation based on radially symmetric concentration profiles $\varphi(r)$ in two dimensions. Since critical profiles correspond to stationary points of the free energy $\tilde{F}$ given by Eqs. (3.66) and (3.45), we next determine **minimal free energy paths** between the homogeneous state and large droplets [65]. We use a measure for the mass concentrated in the nucleus, $N[\varphi] = \int \tanh(w(\varphi - 1/2))\mathrm{d}V$ with $w = 10$, as a reaction coordinate and determine the profile $\varphi(r)$ that minimizes $\tilde{F}$ for a given value N_0 of the constraint using a Lagrange multiplier λ. We thus minimize the constrained free energy

$$\tilde{F}_\lambda[\varphi, \lambda] = \tilde{F}[\varphi] - \lambda(N[\varphi] - N_0) \tag{4.21}$$

by evolving the corresponding partial differential equations

$$\partial_t \varphi = \Lambda_D \nabla^2 \frac{\delta \tilde{F}_\lambda}{\delta \varphi} \tag{4.22a}$$

$$\partial_t \lambda = -\Lambda_L \frac{\delta \tilde{F}_\lambda}{\delta \lambda} , \tag{4.22b}$$

which corresponds to conserved and non-conserved dynamics with mobilities $\Lambda_D = 10^2$ and $\Lambda_L = 10^4$, respectively. Using this procedure, we determine the profile $\varphi(r)$ with Neumann boundary conditions that optimizes $\tilde{F}_\lambda$ for each value N_0 of the constraint, which yields the minimal free energy path.

The profile with the largest free energy $\tilde{F}$ corresponds to the saddle point and thus to the critical nucleus that we sought. The corresponding profiles $\varphi(r)$ are shown and analyzed in Fig. 4.4. Here, the nucleation barrier ΔE is given by the difference of the energy of the critical nucleus to the energy of the homogeneous state.

References

1. Blom K, Ziethen N, Godec A (2022) Thermodynamically consistent phase-field theory including nearest-neighbor pair correlations. Phys Rev Res 5(1):013135. [Online]. Available https://link.aps.org/doi/10.1103/PhysRevResearch.5.013135
2. Cahn JW, Hilliard JE (1958) Free energy of a nonuniform system. I. Interfacial free energy. J Chem Phys 28(2):258–267. [Online]. Available https://aip.scitation.org/doi/10.1063/1.1744102
3. Cahn JW (1959) Free energy of a nonuniform system. II. Thermodynamic basis. J Chem Phys 30(5):1121–1124. [Online]. Available https://aip.scitation.org/doi/10.1063/1.1730145
4. Cahn JW, Hilliard JE (1959) Free energy of a nonuniform system. III. Nucleation in a two-component incompressible fluid. J Chem Phys 31(3):688–699. [Online]. Available https://aip.scitation.org/doi/10.1063/1.1730447

5. Weber CA, Zwicker D, Jülicher F et al (2019) Physics of active emulsions. Rep Prog Phys 82(6):064601. [Online]. Available https://doi.org/10.1088/1361-6633/ab052b
6. Rowlinson JS, Widom B (2013) Molecular theory of capillarity. Courier Corporation
7. Bray AJ (2002) Theory of phase-ordering kinetics. Adv Phys 51(2):481–587. [Online]. Available https://doi.org/10.1080/00018730110117433
8. de Gennes PG (1980) Dynamics of fluctuations and spinodal decomposition in polymer blends. J Chem Phys 72(9):4756–4763. [Online]. Available https://doi.org/10.1063/1.439809
9. Pincus P (1981) Dynamics of fluctuations and spinodal decomposition in polymer blends. II. J Chem Phys 75(4):1996–2000. [Online]. Available https://doi.org/10.1063/1.442226
10. Binder K (1983) Collective diffusion, nucleation, and spinodal decomposition in polymer mixtures. J Chem Phys 79(12):6387–6409. [Online]. Available https://doi.org/10.1063/1.445747
11. Tang Q, Müller M (2021) Evaporation-induced liquid expansion and bubble formation in binary mixtures. Phys Rev Lett 126:028003. [Online]. Available https://link.aps.org/doi/10.1103/PhysRevLett.126.028003
12. Kirschbaum J, Zwicker D (2021) Controlling biomolecular condensates via chemical reactions. J R Soc Interface 18(179):20210255. [Online]. Available https://royalsocietypublishing.org/doi/abs/10.1098/rsif.2021.0255
13. Bo S, Hubatsch L, Bauermann J et al (2021) Stochastic dynamics of single molecules across phase boundaries. Phys Rev Res 3:043150. [Online]. Available https://link.aps.org/doi/10.1103/PhysRevResearch.3.043150
14. Mao S, Chakraverti-Wuerthwein MS, Gaudio H et al (2020) Designing the morphology of separated phases in multicomponent liquid mixtures. Phys Rev Lett 125:218003. [Online]. Available https://link.aps.org/doi/10.1103/PhysRevLett.125.218003
15. Aubert G, Kornprobst P, Aubert G (2006) Mathematical problems in image processing: partial differential equations and the calculus of variations, vol 147. Springer. [Online]. Available https://link.springer.com/book/10.1007/978-0-387-44588-5
16. Copetti MIM, Elliott CM (1992) Numerical analysis of the Cahn-Hilliard equation with a logarithmic free energy. Numerische Mathematik 63(1):39–65. [Online]. Available https://doi.org/10.1007/BF01385847
17. Gómez H, Calo VM, Bazilevs Y et al (2008) Isogeometric analysis of the Cahn–Hilliard phase-field model. Comput Methods Appl Mech Eng 197(49–50):4333–4352. [Online]. Available https://doi.org/10.1016/j.cma.2008.05.003
18. Adamson AW, Gast AP et al (1967) Physical chemistry of surfaces, vol 150. Interscience Publishers, New York
19. Zhao Y, Stein P, Xu B-X (2015) Isogeometric analysis of mechanically coupled Cahn–Hilliard phase segregation in hyperelastic electrodes of Li-ion batteries. Comput Methods Appl Mech Eng 297:325–347. [Online]. Available https://www.sciencedirect.com/science/article/pii/S0045782515003023
20. Zhao Y, Stein P, Bai Y et al (2019) A review on modeling of electro-chemo-mechanics in lithium-ion batteries. J Power Sour 413:259–283. [Online]. Available https://doi.org/10.1016/j.jpowsour.2018.12.011
21. Yang AJM, Fleming PD, Gibbs JH (1976) Molecular theory of surface tension. J Chem Phys 64(9):3732–3747. [Online]. Available https://doi.org/10.1063/1.432687
22. Erlebacher J, Aziz MJ, Karma A et al (2001) Evolution of nanoporosity in dealloying. Nature 410(6827):450–453. [Online]. Available https://doi.org/10.1038/35068529
23. Chen L-Q (2002) Phase-field models for microstructure evolution. Ann Rev Mater Sci 32(1):113–140. [Online]. Available https://doi.org/10.1146/annurev.matsci.32.112001.132041
24. Steinbach I (2009) Phase-field models in materials science. Model Simul Mat Sci Eng 17(7):073001. [Online]. Available https://doi.org/10.1088/0965-0393/17/7/073001
25. Murray JD (2002) Mathematical biology I. An introduction. Springer. [Online]. Available https://link.springer.com/book/10.1007/b98868
26. Sanders DW, Kedersha N, Lee DS et al (2020) Competing protein-RNA interaction networks control multiphase intracellular organization. Cell 181(2):306–324. [Online]. Available https://doi.org/10.1016/j.cell.2020.03.050

27. Hur W, Kemp Jr JP, Tarzia M et al (2020) Cdk-regulated phase separation seeded by histone genes ensures precise growth and function of histone locus bodies. Dev Cell 54(3):379–394. [Online]. Available https://doi.org/10.1016/j.devcel.2020.06.003
28. Spoelstra WK, Jacques JM, Gonzalez-Linares R et al (2021) Crispr-based DNA and RNA detection with liquid-liquid phase separation. Biophys J 120(7):1198–1209. [Online]. Available https://doi.org/10.1016/j.bpj.2021.02.013
29. Stroud D (1973) Simple mean-field approach to phase separation in liquid-metal alloys. Phys Rev B 8:1308–1315. [Online]. Available https://link.aps.org/doi/10.1103/PhysRevB.8.1308
30. Cahn JW (1965) Phase separation by spinodal decomposition in isotropic systems. J Chem Phys 42(1):93–99. [Online]. Available https://doi.org/10.1063/1.1695731
31. Bergmann F, Rapp L, Zimmermann W (2018) Active phase separation: a universal approach. Phys Rev E 98(2):020603. [Online]. Available https://link.aps.org/doi/10.1103/PhysRevE.98.020603
32. Stenhammar J, Tiribocchi A, Allen RJ et al (2013) Continuum theory of phase separation kinetics for active Brownian particles. Phys Rev Lett 111:145702. [Online]. Available https://link.aps.org/doi/10.1103/PhysRevLett.111.145702
33. Safran SA (2018) Statistical thermodynamics of surfaces, interfaces, and membranes. CRC Press
34. Bragg WL, Williams EJ (1934) The effect of thermal agitation on atomic arrangement in alloys. Proc Math Phys Eng Sci 145(855):699–730. [Online]. Available https://royalsocietypublishing.org/doi/abs/10.1098/rspa.1934.0132
35. Flory PJ (1953) Principles of polymer chemistry. Cornell University Press
36. Flory PJ (1942) Thermodynamics of high polymer solutions. J Chem Phys 10(1):51–61. [Online]. Available https://doi.org/10.1063/1.1723621
37. Jacobs WM (2021) Self-assembly of biomolecular condensates with shared components. Phys Rev Lett 126:258101. [Online]. Available https://link.aps.org/doi/10.1103/PhysRevLett.126.258101
38. Wei X, Zhou J, Wang Y et al (2020) Modeling elastically mediated liquid-liquid phase separation. Phys Rev Lett 125:268001. [Online]. Available https://link.aps.org/doi/10.1103/PhysRevLett.125.268001
39. Tripathi K, Menon GI (2019) Chromatin compaction, auxeticity, and the epigenetic landscape of stem cells. Phys Rev X 9:041020. [Online]. Available https://link.aps.org/doi/10.1103/PhysRevX.9.041020
40. Zhang Y, Xu B, Weiner BG et al (2021) Decoding the physical principles of two-component biomolecular phase separation. eLife 10:e62403. [Online]. Available https://doi.org/10.7554/eLife.62403
41. Meibohm J, Esposito M (2022) Finite-time dynamical phase transition in nonequilibrium relaxation. Phys Rev Lett 128:110603. [Online]. Available https://link.aps.org/doi/10.1103/PhysRevLett.128.110603
42. Parry A, Rascón C (2019) The goldstone mode and resonances in the fluid interfacial region. Nat Phys 15(3):287–292. [Online]. Available https://doi.org/10.1038/s41567-018-0361-z
43. Münster G, Cañizares Guerrero M (2021) Interface roughening in two dimensions. J Stat Phys 182(3):1–8. [Online]. Available https://doi.org/10.1007/s10955-021-02738-w
44. Binder K, Evans R, Landau DP et al (1996) Interface localization transition in Ising films with competing walls: Ginzburg criterion and crossover scaling. Phys Rev E 53:5023–5034. [Online]. Available https://link.aps.org/doi/10.1103/PhysRevE.53.5023
45. Albano E, Binder K, Heermann D et al (1990) Critical wetting in the square Ising model with a boundary field. J Stat Phys 61(1):161–178. [Online]. Available https://doi.org/10.1007/BF01013958
46. Rogiers J, Indekeu J (1993) Effect of gravity and confinement on phase equilibria. EPL 24(1):21. [Online]. Available https://doi.org/10.1209/0295-5075/24/1/004
47. Bilalbegović G (1988) Interface delocalization in the three-dimensional Ising model. J Stat Phys 50(5):1131–1140. [Online]. Available https://doi.org/10.1007/BF01019157

48. Binder K, Landau DP, Ferrenberg AM (1995) Character of the phase transition in thin Ising films with competing walls. Phys Rev Lett 74:298–301. [Online]. Available https://link.aps.org/doi/10.1103/PhysRevLett.74.298
49. Binder K, Müller M, Schmid F et al (1999) Interfacial profiles between coexisting phases in thin films: Cahn–Hilliard treatment versus capillary waves. J Stat Phys 95(5):1045–1068. [Online]. Available https://doi.org/10.1023/A:1004510702716
50. Ciach A (1986) Correlation functions in the solid-on-solid model of the fluctuating interface. Phys Rev B 34:1932–1936. [Online]. Available https://link.aps.org/doi/10.1103/PhysRevB.34.1932
51. Stecki J (1993) Capillary length of a planar interface from low temperatures to the critical point: an Ising d = 2 strip. Phys Rev B 47:7519–7524. [Online]. Available https://link.aps.org/doi/10.1103/PhysRevB.47.7519
52. Stecki J, Maciolek A, Olaussen K (1994) Magnetization profiles of the planar fluctuating interface in a d = 2 Ising strip. Phys Rev B 49:1092–1103. [Online]. Available https://link.aps.org/doi/10.1103/PhysRevB.49.1092
53. Ciach A, Stecki J (1987) Scaling in the solid-on-solid interfaces. J Phys A Math Theor 20(16):5619. [Online]. Available https://doi.org/10.1088/0305-4470/20/16/035
54. Albano E, Binder K, Paul W (2000) Monte Carlo studies of d = 2 Ising strips with long-range boundary fields. J Phys Condens Matter 12(12):2701. [Online]. Available https://doi.org/10.1088/0953-8984/12/12/311
55. Fisher ME (1984) Walks, walls, wetting, and melting. J Stat Phys 34(5):667–729. [Online]. Available https://doi.org/10.1007/BF01009436
56. Lauber N, Tichacek O, Bose R et al (2022) Statistical mechanics of biomolecular condensates via cavity methods. arXiv:2201.11581. [Online]. Available https://doi.org/10.48550/arXiv.2201.11581
57. Lin Y-H, Forman-Kay JD, Chan HS (2018) Theories for sequence-dependent phase behaviors of biomolecular condensates. Biochemistry 57(17):2499–2508. [Online]. Available https://doi.org/10.1021/acs.biochem.8b00058
58. Lin Y-H, Forman-Kay JD, Chan HS (2016) Sequence-specific polyampholyte phase separation in membraneless organelles. Phys Rev Lett 117(17):178101. [Online]. Available https://link.aps.org/doi/10.1103/PhysRevLett.117.178101
59. Müller M, MacDowell LG (2000) Interface and surface properties of short polymers in solution: Monte Carlo simulations and self-consistent field theory. Macromolecules 33(10):3902–3923. [Online]. Available https://doi.org/10.1021/ma991796t
60. Squarcini A, Tinti A (2021) Correlations and structure of interfaces in the Ising model: theory and numerics. J Stat Mech Theor Exp 2021(8):083209. [Online]. Available https://doi.org/10.1088/1742-5468/ac1407
61. Guggenheim EA (1935) The statistical mechanics of regular solutions. Proc Math Phys Eng Sci 148(864):304–312. [Online]. Available https://royalsocietypublishing.org/doi/abs/10.1098/rspa.1935.0020
62. Bethe HA (1935) Statistical theory of superlattices. Proc Math Phys Eng Sci 150(871):552–575. [Online]. Available https://royalsocietypublishing.org/doi/abs/10.1098/rspa.1935.0122
63. Kikuchi R, Cahn JW (1962) Theory of domain walls in ordered structures—II: pair approximation for nonzero temperatures. J Phys Chem Solids 23(1–2):137–151. [Online]. Available https://doi.org/10.1016/0022-3697(62)90067-7
64. Parlange J (1968) Phase transition and surface tension in the Guasichemical approximation. J Chem Phys 48(1):169–173. [Online]. Available https://doi.org/10.1063/1.1667895
65. Zwicker D (2020) py-pde: A python package for solving partial differential equations. J Open Sour Softw 5(48):2158. [Online]. Available https://joss.theoj.org/papers/10.21105/joss.02158
66. Fisher ME (1986) Interface wandering in adsorbed and bulk phases, pure and impure. J Chem Soc Faraday Trans II 82(10):1569–1603. [Online]. Available http://dx.doi.org/10.1039/F29868201569
67. Schmitz F, Virnau P, Binder K (2014) Logarithmic finite-size effects on interfacial free energies: phenomenological theory and Monte Carlo studies. Phys Rev E 90:012128. [Online]. Available https://link.aps.org/doi/10.1103/PhysRevE.90.012128

68. Lipowsky R, Kroll DM, Zia RKP (1983) Effective field theory for interface delocalization transitions. Phys Rev B 27:4499–4502. [Online]. Available https://link.aps.org/doi/10.1103/PhysRevB.27.4499
69. Kramers HA (1940) Brownian motion in a field of force and the diffusion model of chemical reactions. Physica 7(4):284–304. [Online]. Available https://www.sciencedirect.com/science/article/pii/S0031891440900982
70. Langer J (1969) Statistical theory of the decay of metastable states. Ann Phys 54(2):258–275. [Online]. Available https://www.sciencedirect.com/science/article/pii/0003491669901535
71. Hänggi P, Talkner P, Borkovec M (1990) Reaction-rate theory: fifty years after Kramers. Rev Mod Phys 62:251–341. [Online]. Available https://link.aps.org/doi/10.1103/RevModPhys.62.251
72. Hartich D, Godec A (2018) Duality between relaxation and first passage in reversible Markov dynamics: rugged energy landscapes disentangled. New J Phys 20(11):112002. [Online]. Available http://dx.doi.org/10.1088/1367-2630/aaf038
73. Yang CN (1952) The spontaneous magnetization of a two-dimensional Ising model. Phys Rev 85:808–816. [Online]. Available https://link.aps.org/doi/10.1103/PhysRev.85.808
74. Spirin V, Krapivsky PL, Redner S (2001) Fate of zero-temperature Ising ferromagnets. Phys Rev E 63:036118. [Online]. Available https://link.aps.org/doi/10.1103/PhysRevE.63.036118
75. Barros K, Krapivsky PL, Redner S (2009) Freezing into stripe states in two-dimensional ferromagnets and crossing probabilities in critical percolation. Phys Rev E 80:040101. [Online]. Available https://link.aps.org/doi/10.1103/PhysRevE.80.040101
76. Metropolis N, Rosenbluth AW, Rosenbluth MN et al (1953) Equation of state calculations by fast computing machines. J Chem Phys 21(6):1087–1092. [Online]. Available https://doi.org/10.1063/1.1699114
77. Abraham DB (1981) Capillary waves and surface tension: an exactly solvable model. Phys Rev Lett 47:545–548. [Online]. Available https://link.aps.org/doi/10.1103/PhysRevLett.47.545
78. Müller M, Münster G (2005) Profile and width of rough interfaces. J Stat Phys 118(3):669–686. [Online]. Available https://doi.org/10.1007/s10955-004-8824-2
79. Fisher MPA, Fisher DS, Weeks JD (1982) Agreement of capillary-wave theory with exact results for the interface profile of the two-dimensional Ising model. Phys Rev Lett 48:368–368. [Online]. Available https://link.aps.org/doi/10.1103/PhysRevLett.48.368
80. Wang X (2009) Asymptotics of the q-theta function. Commun Math Anal 7:50–54. [Online]. Available https://userweb.ucs.louisiana.edu/~xxw6637/papers/CMA2009.pdf
81. Mukhopadhyay A, Law BM (2000) Casimir effect in critical films of binary liquid mixtures. Phys Rev E 62:5201–5215. [Online]. Available https://link.aps.org/doi/10.1103/PhysRevE.62.5201
82. Dantchev DM, Dietrich S (2022) Critical Casimir effect: exact results. arXiv:2203.15050. [Online]. Available https://arxiv.org/abs/2203.15050

Chapter 5
Criticality in Cell Adhesion

It ain't what you don't know that gets you into trouble. It's what you know for sure that just ain't so.

Mark Twain in The Big Short

In this chapter we apply the Ising model to describe the statics and kinetics of cell adhesion. We illuminate the many-body effects underlying the structure, formation, and dissolution of cellular adhesion domains in the presence and absence of forces. We consider mixed **Glauber-Kawasaki dynamics** of a two-dimensional model of **nearest-neighbor interacting diffusing adhesion bonds** with intrinsic binding-affinity under the action of a shared pulling or pushing force. Accurate analytical results are obtained by employing the BG approximation for the thermodynamics and kinetics of adhesion clusters of any size, including the thermodynamic limit. A new kind of **dynamical phase transition** is uncovered—the mean formation and dissolution times per adhesion bond change discontinuously with respect to the bond-coupling parameter. At the respective critical points cluster formation and dissolution are fastest, while the statistically dominant transition path undergoes a qualitative change. In the context of the Ising model the dynamical phase transition reflects a first-order discontinuity in the magnetization-reversal time. Our results provide a potential explanation for the mechanical regulation of cell adhesion, and suggest that the quasi-static and kinetic response to changes in the membrane stiffness or applied forces is largest near the statical and dynamical critical point, respectively. Part of this chapter has been published in *Physical Review X* under the terms of the Creative Commons Attribution 4.0 International license [1]. Analytical results obtained with the MF approximation are omitted from this chapter and are referred to in [1] and Sect. 2.10.

K. Blom, *Pair-Correlation Effects in Many-Body Systems*, Springer Theses,
https://doi.org/10.1007/978-3-031-29612-3_5

5.1 Introduction

Most of our current understanding on the formation and stability of adhesion clusters derives from the analysis of individual [2] and non-interacting adhesion bonds [3–5], and studies of collective effects in biomimetic vesicular model systems with floppy membranes [6, 7] and mobile cellular adhesion molecules CAMs [8]. These results therefore do not necessarily apply to cells, where membranes are stiffened by the presence of, and receptors are anchored to, the stiff actin cytoskeleton that can actively exert forces on the membrane [9].

Diverse aspects of biological adhesion have been investigated experimentally by contact-area fluorescence recovery after photobleaching [10], Förster resonance energy transfer [11], metal-induced energy transfer [12], reflection interference contrast microscopy [13], optical tweezers [14], flow-chamber methods [15, 16], centrifugation assays [17, 18], biomembrane force probe [19, 20], micropipette techniques [21, 22], and atomic force microscopy (AFM) [23–29] (see Fig. 5.1 for some examples). Experiments unraveled a collective behavior of clusters of adhesion bonds that cannot be explained as a sum of their individual behavior [11, 30–33] that is meanwhile well understood (see e.g. [34, 35]). More specifically, the opening/closing of adhesion bonds is profoundly affected by membrane fluctuations even if their amplitude becomes as small as 0.5 nm—smaller than the thickness of the membrane itself [36, 37].

These observations imply many-body physics to be at play, i.e. an interplay between the coupling of nearby adhesion bonds through deformations of the fluctuating membrane and mechanical forces acting on the membrane [7, 32, 33, 36–45]. Supporting the idea are experimental observations of cells changing the membrane flexibility and/or membrane fluctuations through ATP-driven activity [46–49], decoupling the F-actin network [50] or remodelling the actomyosin cytoskeleton [49], and through acidosis [37], in order to alter adhesion binding rates and strength [33, 37, 51–54] or to become motile [55]. There is also a striking correspondence between membrane stiffness and the metastatic potential of cancer cells—the stiffness of cancer cells was found to determine their migration and invasion potential [55]. The effect is not limited to cells; the elastic modulus was similarly found to significantly affect the specific adhesion of polymeric networks [56].

Notwithstanding all theoretical efforts [35, 36, 38–44], a consistent and comprehensive physical picture of **collective adhesion** under the action of a mechanical force that could explain the observations on live cellular systems [32, 33, 51–55, 57] remains elusive. For example, whether the coupling of individual bonds causes the collective association and dissociation rates to increase or decrease, respectively, was speculated to depend on the intrinsic single-bond affinity [11, 58], cell type (i.e. surface corrugation) [30] and on the state of the actin cytoskeleton [11]. An understanding of cellular adhesion therefore must integrate the complex interplay between the correlated, collective (un)binding [6, 7, 33, 40, 42–44], the intrinsic affinity of anchored adhesion bonds [11, 58, 59], the cell type and surface topology

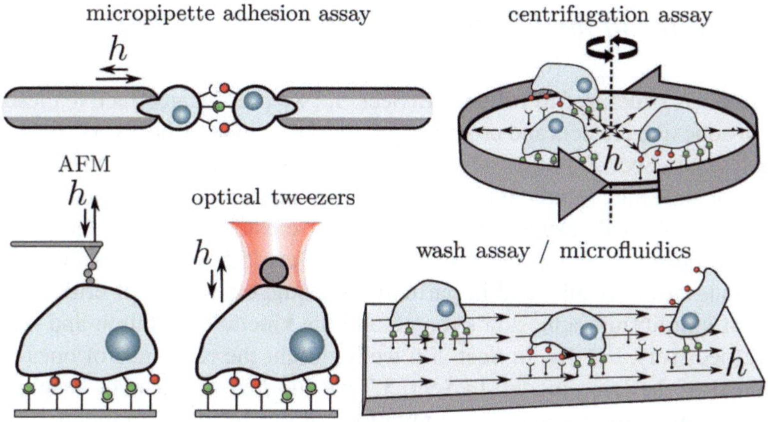

Fig. 5.1 Examples of experimental techniques to measure adhesion strength. For each technique an external force h is applied to a cell (blue) which is attached to a substrate or other cell through the cellular adhesion molecules (red: detached, green: attached)

[30], as well as the integrity of, and forces generated by, the actin cytoskeleton [11, 25, 52–54] under physiological [57] or pathological conditions [55, 60–62].

Although cellular adhesion is omnipresent in various biological systems with different underlying microsopic details, we attempt to capture the essential conclusions by incorporating the most general and essential features of adhesion molecules and their dynamics. In order to arrive at a deeper understanding of the mechanical regulation of cellular adhesion that would explain the collective dynamics of adhesion bonds on the level of individual (un)binding events, we here consider mixed Glauber-Kawasaki dynamics of a generic, two-dimensional model of diffusing nearest-neighbor interacting adhesion bonds with intrinsic affinity μ under the action of a shared force h (see Fig. 5.2a).

Highly accurate analytical results on the BG level reveal the many-body—that is beyond MF—physics underlying biological adhesion. We consider in detail cluster-sizes ranging from a few CAMs to the thermodynamic limit. In the thermodynamic limit we determine the equation of state and complete phase behavior that displays phase separation and co-existence of dense and dilute adhesion domains. The critical behavior is investigated in detail and striking differences are found between pulling- and pushing-forces. Strikingly, we prove the existence of a seemingly new kind of *dynamical phase transition*—the mean first passage time to cluster formation/dissolution is proven to **change discontinuously with respect to the coupling strength**. This dynamical phase transition, and more generally the non-linear and non-monotonic dependence on the membrane flexibility, may explain the puzzling cooperative behavior of effective (un)binding rates measured experimentally.

5.1.1 Outline of the Chapter

This chapter is structured as follows: In Sect. 5.2 we present an effective mesoscopic model of adhesion clusters and provide a practical roadmap to the diverse calculations and analyses. In Sect. 5.3 we present explicit analytical results for the thermodynamic equation of state and complete phase behavior of adhesion clusters, and in Sect. 5.4 we present analytical results for the kinetics of cluster formation and dissolution both in the presence and absence of forces. In Sect. 5.5 we discuss the biological implications of our results and in particular the suggestive rôle of criticality in the context of equilibrium adhesion strength and the kinetic dissolution and formation rates, respectively. Finally, in Sect. 5.6 we highlight the relevance of our results in the context of the Ising model. In Sect. 5.7 we summarize our main results, give a perspective on the importance and limitations of our results, and mention possible extensions to be made in future studies. Details of calculations, explicit asymptotic results, and further technical information is presented in a series of Appendices and in [1].

5.2 Interacting Adhesion Bonds Under Shared Force

5.2.1 Equilibrium

We consider a two-dimensional patch of a cell surface with N adhesion molecules embedded in the cell membrane, their lateral positions forming a lattice with coordination number $\bar{z}$ (see Fig. 5.2). The results we derive hold for any lattice but we focus the discussion mainly on the square lattice ($\bar{z} = 4$) with free boundary conditions. Opposing the patch is a stiff substrate or a neighboring cell-patch with complementary adhesion molecules occupying a commensurate lattice. The state of individual bonds is denoted by σ_i with $i = 1, 2, \ldots, N$, where $\sigma_i = +1$ if bond i is broken an $\sigma_i = -1$ if it is closed.

In the presence of a **timescale separation** the opening/closing of nearest neighbor bonds is coupled via membrane fluctuations. Following closely the arguments of Ref. [39], we can integrate out the membrane degrees of freedom to obtain an effective Ising-like model for the bonds within the patch with **effective Hamiltonian**

$$\mathcal{H}(\{\sigma_i\}) = -J \sum_{\langle ij \rangle} \sigma_i \sigma_j - \mu N_c(\{\sigma_i\}) + \mathcal{H}_{\mathrm{h}}(\{\sigma_i\}), \tag{5.1}$$

where $J \geq 0$ is the membrane-induced short-range coupling between the bonds, $\langle ij \rangle$ denotes all nearest-neighbor pairs, μ is the effective chemical potential (i.e. **intrinsic affinity**[1]) of individual closed bonds $N_c(\{\sigma_i\}) \equiv \sum_i \delta_{\sigma_i,-1}$, where δ_{ik} is Kronecker's

[1] The intrinsic affinity μ acts similarly as the external magnetic field h defined in Eq. (1.1).

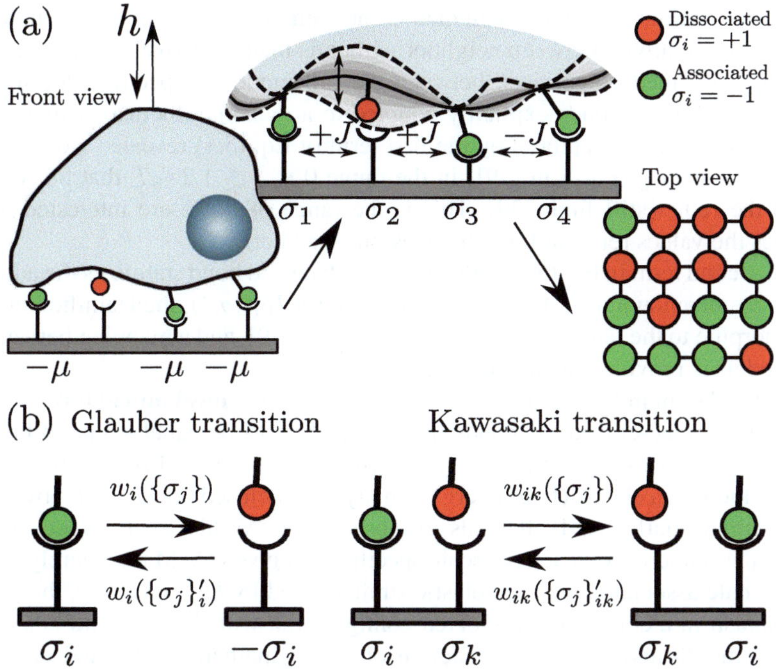

Fig. 5.2 A coarse-grained model for the cooperative association/dissociation of adhesion bonds. **a** Schematic of the effective many-body model governed by Eqs. (5.1) and (5.2), depicting an adhesion domain on a cell-patch with 16 CAMs anchored to stiff substrate. Adhesion bonds are arranged on a 4×4 square lattice and can assume two states, $\sigma_i = \pm 1$, where $+1$ corresponds to an open (red) and -1 to a closed bond (green). Nearest-neighbor bonds experience an effective interaction J induced by undulations of the anchoring membrane. An external force h is pulling/pushing on the adhesion domain. Each adhesion bond has an intrinsic binding-affinity $\mu \geq 0$ that favors a bound state. A small number of bonds is depicted for convenience only. We consider different system sizes including the thermodynamic limit. **b** Glauber and Kawasaki transition. A Glauber transition changes the binding state of a single adhesion bond $\sigma_i \to -\sigma_i$ with transition rate $w_i(\{\sigma_j\})$ [see Eq. (5.3)]. A Kawasaki transition interchanges two nearest-neighbor adhesion bonds $\sigma_i \leftrightarrow \sigma_k$ with transition rate $w_{ik}(\{\sigma_j\})$ [see Eq. (5.5)]

delta, and $\mathcal{H}_{\mathrm{h}}(\{\sigma_i\})$ is the Hamiltonian describing the effect of the mechanical force. The first term in Eq. (5.1) represents the effective coupling between nearest neighbor bonds, and is isomorphic to the interaction term in the Ising model [63]. It is an effective measure of bond-cooperativity, i.e. it reflects that the (free) energy penalty of closing/breaking a bond is smaller if neighboring bonds are closed/open, respectively [39]. Such an effective description in terms of bonds coupled via a short-range membrane-mediated interaction is feasible when bonds are flexible and/or the patch of the cell membrane is quite (but not completely) stiff and is thus rather pulled down as a whole instead of being locally strongly deformed by the binding of individual bonds [39]. In this limit the coupling strength is determined by the effective bending rigidity of the cell membrane, κ, via $J \propto 1/\sqrt{\kappa}$ (see [39] and Appendix 2). That is, in this regime a relatively floppier cell membrane with lower bending rigidity induces a stronger cooperativity between neighboring bonds than a relatively stiff membrane. Notably, a detailed comparison between the full model of specific adhesion (i.e. reversible adhesion bonds explicitly coupled to a dynamic fluctuating membrane) and the lattice model captured by the first term of Eq. (5.1) revealed a quantitative agreement (see e.g. Fig. 5 in [39]) in the range $0 \leq J \lesssim 1.2\, k_{\mathrm{B}}T$ that lies entirely within the rather stiff limit [39]. This is the range of J we are interested in and includes the values relevant for cell adhesion (see Sect. 5.5).

The second term in Eq. (5.1) reflects that each closed bond stabilizes the adhesion cluster by an amount $-\mu$. Aside from the last term $\mathcal{H}_{\mathrm{h}}(\{\sigma_i\})$, the Hamiltonian (5.1) is isomorphic to the lattice gas model developed in [39], and a mapping between the two models is provided in Appendix 2.

The third term in Eq. (5.1), $\mathcal{H}_{\mathrm{h}}(\{\sigma_i\})$, accounts for the **mechanical force** h acting on the membrane-embedded bonds that we assume to be equally shared between all N_c *closed* bonds.[2] More precisely, external force h destabilizes the bound state by introducing an elastic (free) energy penalty on all closed bonds, whereby broken bonds remain unaffected. If all bonds are closed, $N_c = N$, this penalty is set to be hx_0, where x_0 is a microscopic length-scale specific for a given CAM that merely sets the energy scale associated with the elastic strain caused by h. Conversely, the penalty must vanish in a completely dissolved configuration with $N_c = 0$, and is assumed to be a smooth and monotonic function of N_c. A mathematically and physically consistent definition is

$$\mathcal{H}_{\mathrm{h}}(\{\sigma_i\}) = -2hx_0 \left((1 + N_c(\{\sigma_i\})/N)^{-1} - 1 \right). \tag{5.2}$$

A **pulling force**, $h > 0$, favors the dissociation of bonds while a **pushing force**, $h < 0$, favors their association. We are interested in strain energies on the order of the thermal energy per bond, i.e. $|h|x_0/N = \mathcal{O}(k_{\mathrm{B}}T)$. Note that the assumption of an equally shared force in Eq. (5.2) is valid if either of the following conditions is satisfied: the anchoring membrane has a large combined elastic modulus (i.e. stiff membranes or membrane/substrate pairs), individual bonds are flexible, the bond-density is low, or the membrane is prestressed by the actin cytoskeleton [35, 64, 65]. In the limit of a rather stiff membrane both, a spin representation with effective coupling J and a uniform force load are valid approximations to describe cell adhesion under force over a broad range of physically relevant parameters, as we detail below. The implications of a non-uniform force load are addressed in detail in Sect. 5.7 and Appendix 3.

[2] Here h refers to the external force and is **not** equal to the external magnetic field in Eq.(1.1).

5.2.2 Kinetics

The breaking/closure and lateral diffusion of adhesion bonds are assumed to evolve as a **discrete time Markov chain** with mixed **single-bond-flip Glauber dynamics** [66] and **two-bond-exchange Kawasaki dynamics** [67] (see Fig. 5.2b). For a single jump in the Markov chain we define the probability to attempt a Glauber transition as $p_k \in [0, 1]$ which controls the diffusion rate, and for the sake of generality is allowed to depend on the number of closed bonds k. Similarly, the probability to attempt a Kawasaki transition is given by $1 - p_k \in [0, 1]$. Here we consider the scenarios of purely anchored ($p_k = 1, \ \forall k$) and mixed diffusing bonds ($0 < p_k < 1, \ \forall k$), whereas purely diffusing bonds ($p_k = 0, \ \forall k$) will not be considered since bonds break and associate in real biological systems. Further details about the respective transition rates are given below.

Glauber transitions: Let $\{\sigma_j\}_i'$ denote the bond configuration obtained by flipping bond i while keeping the configuration of all other bonds fixed, i.e. $\{\sigma_j\}_i' \equiv (-\sigma_i, \{\sigma_{j\neq i}\})$. Moreover, let $w_i(\{\sigma_j\})$ denote the transition rate from $\{\sigma_j\}$ to $\{\sigma_j\}_i'$ and $\Delta\mathcal{H}_i(\{\sigma_j\}) \equiv \mathcal{H}(\{\sigma_j\}_i') - \mathcal{H}(\{\sigma_j\})$ the energy difference associated with the transition. These rates can be specified uniquely by limiting interactions to nearest-neighbors, imposing isotropy in position space, and requiring that w_i satisfies DB, i.e. $w_i(\{\sigma_j\})/w_i(\{\sigma_j\}_i') = \exp(-\beta\Delta\mathcal{H}_i(\{\sigma_j\}))$, where $\beta = 1/k_\mathrm{B}T$ is the inverse thermal energy. The general result reads $w_i(\{\sigma_j\}) = \alpha[1 - \tanh(\beta\Delta\mathcal{H}_i(\{\sigma_j\})/2)]/2N$, where α is an intrinsic attempt-frequency that sets the fastest timescale [66], and time will throughout be expressed in units of α^{-1}. Introducing furthermore the dimensionless quantities $\tilde{J} = \beta J$, $\tilde{\mu} = \beta\mu$ and $\tilde{h} = \beta h x_0/N$, leads to

$$w_i(\{\sigma_j\}) = \frac{\alpha}{2N}\left\{1 - \sigma_i \tanh\left[\tilde{J}\sum_{\langle ij\rangle}\sigma_j - \frac{\tilde{\mu}}{2} + \Lambda^{\tilde{h}}_{\{\sigma_j\},i}\right]\right\}, \tag{5.3}$$

where we defined the auxiliary function

$$\Lambda^{\tilde{h}}_{\{\sigma_j\},i} \equiv \frac{\tilde{h}}{(1 + N_c(\{\sigma_j\})/N)(1 + N_c(\{\sigma_j\}_i')/N)}. \tag{5.4}$$

Kawasaki transitions: Let $\{\sigma_j\}_{ik}'$ denote the bond configuration upon interchanging the state of the nearest neighbor bonds σ_i and σ_k while keeping the configuration of all other bonds fixed, i.e. $\{\sigma_j\}_{ik}' \equiv (\sigma_i \leftrightarrow \sigma_k, \{\sigma_{j\neq(i,k)}\})$. We denote the Kawasaki transition rate from $\{\sigma_j\}$ to $\{\sigma_j\}_{ik}'$ as $w_{ik}(\{\sigma_j\})$, where $\Delta\mathcal{H}_{ik}(\{\sigma_j\}) \equiv \mathcal{H}(\{\sigma_j\}_{ik}') - \mathcal{H}(\{\sigma_j\})$ is the energy difference associated with the transition. Imposing the same symmetry constraints as for the Glauber rates, as well as DB, yields the general expression [67]

$$w_{ik}(\{\sigma_j\}) = \frac{\alpha}{2N}\left\{1 - \frac{\sigma_i - \sigma_k}{2}\tanh\left[\tilde{J}(\sum_{\langle ij\rangle}\sigma_j - \sum_{\langle kl\rangle}\sigma_l)\right]\right\}, \tag{5.5}$$

where we have used that $(\sigma_i - \sigma_k)/2 \in \{-1, 0, 1\}$. As pointed out in [67], the transition is only meaningful when $\sigma_k = -\sigma_i$, otherwise the transition brings the system to an identical state, which is equivalent to no transition. Note that the Kawasaki rates given by Eq. (5.5) do not depend on the external force $\tilde{h}$ nor the binding-affinity $\tilde{\mu}$, since the Kawasaki transition conserves the total number of open and closed adhesion bonds. However, if we introduce a position-dependent force/binding affinity, the Kawasaki rates also depend on $\tilde{h}$ and $\tilde{\mu}$, which we analyze in Appendix 3.

5.2.3 Strategy Roadmap

We focus in detail on both, the equilibrium properties as well as the kinetics of cluster formation and dissolution for all cluster sizes. A roadmap to our extensive analysis is presented in Fig. 5.3. For small to moderate cluster sizes, i.e. up to 50 bonds for the equilibrium properties and up to 25 bonds in the case of formation/dissolution kinetics, we obtain exact solutions using standard algebraic methods [68]. To circumvent the explosion of combinatorial complexity for large system sizes, we employ the BG approximation (see Chap. 2) to derive closed-form expressions for the partition function, and finally carry out the thermodynamic limit to derive explicit closed-form results for large adhesion clusters. When considering the formation/dissolution kinetics of large clusters and in the thermodynamic limit, we apply the **local equi-**

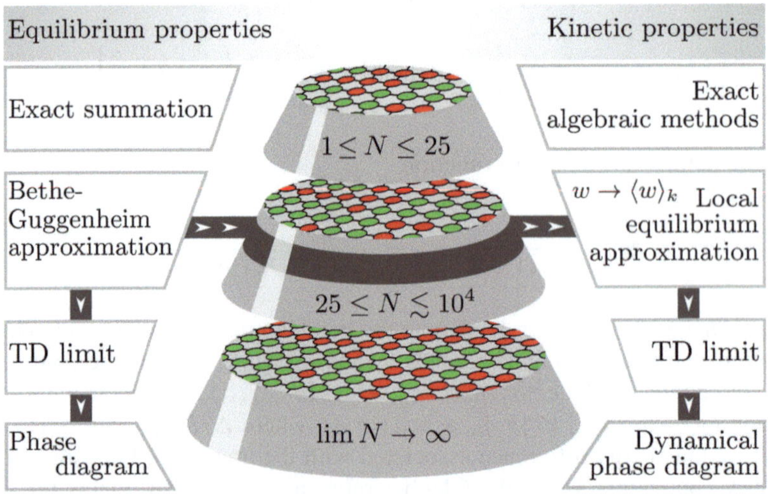

Fig. 5.3 Strategy roadmap. Small system sizes $N \leq 5 \times 5$ are solved for exactly. The thermodynamics of larger systems is treated on the level of the BG approximation and the kinetics by assuming *local equilibrium*. Within the BG approximation we take the thermodynamic limit $N \to \infty$ and determine the phase behavior, master scaling of dissolution/formation kinetics, and analyze the statical and dynamical critical behavior

librium approximation (LEQ), where we assume that the growth and dissolution evolves like a birth-death process on the free energy landscape.

5.3 Equilibrium Properties of Adhesion Clusters

5.3.1 Small and Intermediate Clusters

In order to quantify the equilibrium stability of adhesion clusters, we first analyze the equation of state (EOS) for the average fraction of closed bonds, $\langle\varphi\rangle \equiv \langle N_c(\{\sigma_i\})\rangle/N$ at given $\tilde{\mu}$, $\tilde{J}$ and $\tilde{h}$. To this end we require $\mathcal{Z}_k$, the partition function constrained to the number of closed bonds $N_c(\{\sigma_i\}) = k$. We therefore write the total canonical partition function $\mathcal{Z}$ for a system of N adhesion bonds as $\mathcal{Z} \equiv \sum_{\{\sigma_i\}} \mathrm{e}^{-\beta\mathcal{H}(\{\sigma_i\})} \equiv \sum_{k=0}^{N} \hat{\mathcal{Z}}_k$, where

$$\hat{\mathcal{Z}}_k \equiv \sum_{\{\sigma_i\}} \mathrm{e}^{-\beta\mathcal{H}(\{\sigma_i\})}\delta_{N_c(\{\sigma_i\}),k} = \mathrm{e}^{[\tilde{\mu}-2\tilde{h}(k/N+1)^{-1}]k}\mathcal{Z}_k, \tag{5.6}$$

and $\mathcal{Z}_k \equiv \sum_{\{\sigma_i\}} \exp(\tilde{J}\sum_{\langle ij\rangle}\sigma_i\sigma_j)\delta_{N_c(\{\sigma_i\}),k}$ is the partition function of the Ising model at zero field conditioned to have a magnetization $N/2 - k$ (see also Sect. 1.1.5). The free energy density (per bond) in units of thermal energy $k_\mathrm{B}T$ constrained to a given fraction of closed bonds φ, $\tilde{f}_N(\varphi)$, and the EOS, $\langle\varphi(\tilde{\mu}, \tilde{J}, \tilde{h})\rangle$, are given by

$$\tilde{f}_N(\varphi) = -N^{-1}\ln\hat{\mathcal{Z}}_k, \quad \langle\varphi\rangle = N^{-1}\partial_{\tilde{\mu}}\ln\mathcal{Z}. \tag{5.7}$$

We note that $\mathrm{e}^{-N\tilde{f}_N(\varphi)}/\mathcal{Z} = \mathrm{Prob}(N_c = N\varphi)$ in an equilibrium ensemble of N bonds. The sum over constrained configurations in $\mathcal{Z}_k$ contains $\binom{N}{k}$ terms. Whereas it can be performed exactly for $N \lesssim 50$, it explodes for larger system sizes. To overcome the computational complexity, we employ the BG approximation [69], yielding (see derivation in Chap. 2)

$$\mathcal{Z}_{k=N\varphi} \approx \mathcal{Z}^{\mathrm{BG}}_{k=N\varphi} = \binom{N}{N\varphi}\frac{\Psi_{\mathrm{BG}}(\varphi, \zeta^\dagger_{\mathrm{BG}}(\varphi))}{\Psi_{\mathrm{BG}}(\varphi, \zeta_{\mathrm{BG}}(\varphi))}\mathrm{e}^{-2\bar{z}N\tilde{J}(\zeta^\dagger_{\mathrm{BG}}(\varphi)-1/4)}. \tag{5.8}$$

where the functions Ψ_{BG} and $\zeta^\dagger_{\mathrm{BG}}$ are given by Eqs. (2.9) and (2.12), respectively.

Figure 5.4a–c shows a comparison of the free energy density $\tilde{f}_N(\varphi)$ for a cluster of 40 bonds for various affinities $\tilde{\mu}$ and external forces $\tilde{h}$, and confirms the high accuracy of the BG approximation on the one hand, and the systematic failure of the MF result on the other hand. This signifies that correlations between adhesion bonds decisively affect cluster properties. Moreover, pairwise correlations captured by the BG approach are apparently dominant, whereas higher order correlations that were ignored are apparently insignificant.

Similarly, in Fig. 5.4d–f we depict the EOS for a cluster of 40 bonds. The BG approximation (blue lines) is very accurate for all values of $\tilde{J}$, whereas the MF

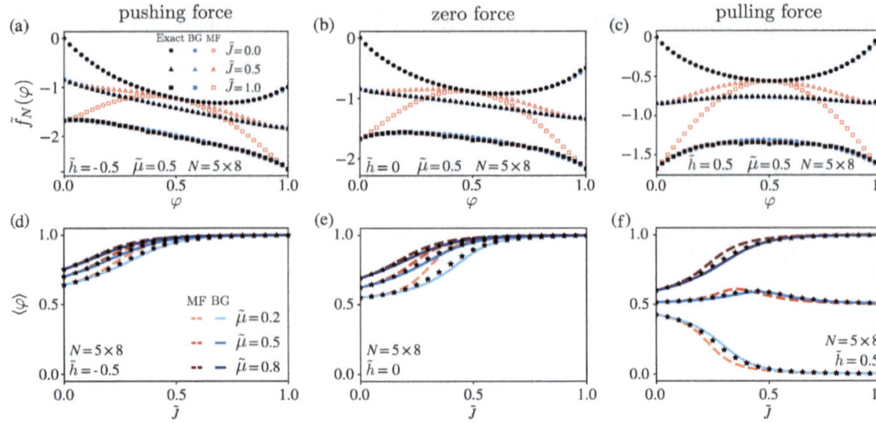

Fig. 5.4 **Free energy landscape and EOS for small clusters.** **a–c** Free energy density conditioned on φ, $\tilde{f}_N(\varphi)$ from Eq. (5.7) for a system of $N = 5 \times 8$ bonds on a square lattice for **a** a pushing force $\tilde{h} = -0.5$, **b** no force $\tilde{h} = 0$, and **c** pulling force $\tilde{h} = 0.5$; Black symbols depict exact results, blue symbols the BG approximation, and red symbols the MF result. **d–f** EOS, $\langle\varphi\rangle$, for a cluster of 5×8 adhesion bonds on a square lattice as a function of the dimensionless coupling $\tilde{J}$ for **d** a pushing force $\tilde{h} = -0.5$, **e** no force $\tilde{h} = 0$, and **f** a pulling force $\tilde{h} = 0.5$. Symbols depict exact results, blue lines correspond to the BG approximation, and red to the MF result

approximation (red lines) fails for intermediate values of the coupling. We observe striking differences in the dependence of $\langle\varphi\rangle$ on the coupling $\tilde{J}$ (and hence membrane rigidity) with respect to the intrinsic binding-affinity $\tilde{\mu}$ in the presence of a pulling force (see Fig. 5.4f). At strong coupling between adhesion bonds $\langle\varphi\rangle$ depends strongly on $\tilde{\mu}$. In the presence of a pulling force adhesion bonds with a weak affinity are on average all broken, whereas they are all closed if the affinity is large. Notably, the dependence of $\langle\varphi\rangle$ on the coupling $\tilde{J}$ at zero force (see Fig. 5.4e) agrees qualitatively well with experimental observations [11, 11, 30, 58] and hints at some form of critical behavior underneath, which we discuss in more detail in Sect. 5.5.

5.3.2 Thermodynamic Limit

To explore the phase diagram in detail and analyze the critical behavior, we consider the thermodynamic (TD) limit of the BG and MF free energy density, i.e. the scaling limit

$$\tilde{\mathrm{f}}_{\mathrm{BG,MF}}(\varphi) \equiv \lim_{k/N=\varphi=\mathrm{const.}}^{N\to\infty}[\tilde{f}_N^{\mathrm{BG,MF}}(\varphi)], \tag{5.9}$$

which exists and is given by $\tilde{\mathrm{f}}_{\mathrm{BG}}(\varphi) = \tilde{\mathrm{u}}_{\mathrm{BG}}(\varphi) - \tilde{\mathrm{s}}_{\mathrm{BG}}(\varphi)$, where the internal energy density and entropy density are respectively given by

$$\tilde{\mathrm{u}}_{\mathrm{BG}}(\varphi) = 2\bar{z}\tilde{J}[\zeta_{\mathrm{BG}}^{\dagger}(\varphi) - 1/4] + 2\tilde{h}\varphi/(1+\varphi) - \tilde{\mu}\varphi,$$
$$\tilde{\mathrm{s}}_{\mathrm{BG}}(\varphi) = (\bar{z}-1)[\Xi(\varphi) + \Xi(1-\varphi)] - (\bar{z}/2)[\Xi(\varphi - \zeta_{\mathrm{BG}}^{\dagger}(\varphi)) + \Xi(1-\varphi-\zeta_{\mathrm{BG}}^{\dagger}(\varphi)) + 2\Xi(\zeta_{\mathrm{BG}}^{\dagger}(\varphi))], \tag{5.10}$$

where $\Xi(x) \equiv x \ln x$ and $\zeta_{\mathrm{BG}}^{\dagger}$ is given by Eq. (2.12). The result for $\tilde{\mathrm{f}}_{\mathrm{MF}}(\varphi)$ is given in Appendix C of [1]. Somewhat surprisingly, the free energy density of a finite system, $\tilde{f}_N^{\mathrm{BG}}(\varphi)$, converges to the thermodynamic limit $\tilde{\mathrm{f}}_{\mathrm{BG}}(\varphi)$ already for $N \gtrsim 100$. For convenience, we henceforth drop the subscript BG when considering the BG result, i.e. $\tilde{\mathrm{f}}_{\mathrm{BG}}(\varphi) \to \tilde{\mathrm{f}}(\varphi)$.

The EOS in the thermodynamic limit is determined by means of the saddle-point method (see Appendix 5), yielding a weighted sum over φ_i^0, the M *global* minima of $\tilde{\mathrm{f}}(\varphi)$:

$$\langle\varphi\rangle_{\mathrm{TD}} = \lim_{k/N=\varphi=\mathrm{const.}}^{N\to\infty}[N^{-1}\partial_{\tilde{\mu}} \ln \mathcal{Z}] \simeq \textstyle\sum_{i=1}^{M} c_i \varphi_i^0, \tag{5.11}$$

where $\tilde{\mathrm{f}}(\varphi_i^0) = \tilde{\mathrm{f}}_{\min}$, $\forall i$, and $\simeq$ stands for asymptotic equality in the thermodynamic limit. In practice, M is either 1 (unique minimum) or 2 (two-fold degenerate minima). The minima have the universal form $\varphi_m^0 = \xi^4_{\tilde{\mu},\tilde{J},\tilde{h}}/(1+\xi^4_{\tilde{\mu},\tilde{J},\tilde{h}})$ with the coefficients $\xi_{\tilde{\mu},\tilde{J},\tilde{h}}$ and weights c_i given explicitly in Appendix 5. The EOS $\langle\varphi\rangle$ for a finite cluster seems to converge to the saddle-point asymptotic $\langle\varphi\rangle_{\mathrm{TD}}$ already for $N \gtrsim 400$ for any value of the force $\tilde{h}$, bond affinity $\tilde{\mu}$, and coupling $\tilde{J}$ (see Fig. 5.5a–c), and is qualitatively the same as for smaller clusters (compare Fig. 5.5a–c with Fig. 5.4d–f). However, important differences emerge in the thermodynamic limit—the system may undergo a phase transition and phase-separate into dense ("liquid") and dilute ("gas") phases of closed bonds with composition φ_l and φ_g, respectively (see also [44]).

5.3.3 Phase Diagram and Critical Behavior

To determine the phase diagram we require the binodal $\tilde{J}_b(\varphi)$ and spinodal $\tilde{J}_s(\varphi)$ line, which are introduced in Sect. 2.8. Recall that the binodal and spinodal are independent of the intrinsic affinity $\tilde{\mu}$ which couples linearly to φ. For a non-zero force ($\tilde{h} \neq 0$) we determine $\tilde{J}_b(\varphi)$ numerically, whereas for $\tilde{h} = 0$ the binodal is given by Eq. (2.20). The spinodal line for any force $\tilde{h}$ is in turn given by

$$\tilde{J}_s(\varphi, \tilde{h}) = \frac{1}{4} \ln \left\{ \frac{[\varphi - \Phi(\varphi, \tilde{h})][1 - \varphi - \Phi(\varphi, \tilde{h})]}{\Phi(\varphi, \tilde{h})^2} \right\}, \tag{5.12}$$

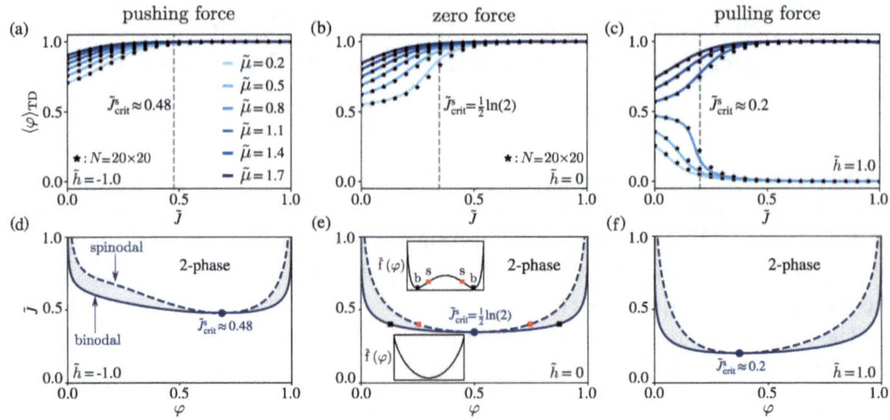

Fig. 5.5 EOS and phase diagram in the thermodynamic limit. **a–c** $\langle\varphi\rangle$ for a cluster of 20×20 adhesion bonds on a square lattice (symbols) and the saddle-point asymptotic $\langle\varphi\rangle_{\rm TD}$ from Eq. (5.11) as a function of the dimensionless coupling $\tilde{J}$ for various affinities $\tilde{\mu}$ and for **a** a pushing force $\tilde{h} = -1$, **b** no force, and **c** a pulling force $\tilde{h} = 1$. The dashed vertical line denotes the (statical) *critical* coupling strength $\tilde{J}^s_{\rm crit}$ whereupon the system phase-separates into dense and dilute phases of closed bonds. **d–f** Phase diagram for **d** a pushing force $\tilde{h} = -1$, **e** no force, and **f** a pulling force $\tilde{h} = 1$; the full and dashed lines depict the binodal and spinodal line, respectively. The shaded area depicts the region where the system is metastable. The blue circle depicts the (statical) critical point $(\varphi^s_{\rm crit}, \tilde{J}^s_{\rm crit})$. Inset in **e**: Schematic of the free energy landscape $\tilde{f}(\varphi)$ below the critical coupling $\tilde{J} < \tilde{J}^s_{\rm crit}$ (bottom) displaying a single minimum, and a bi-stable free energy landscape above the critical coupling $\tilde{J} > \tilde{J}^s_{\rm crit}$ (top), with the black and red symbols illustrating the meaning of phase compositions highlighted in the phase diagram

with the auxiliary function

$$\Phi(\varphi, \tilde{h}) \equiv 2\varphi(1-\varphi) + \bar{z}\left[\frac{1-\bar{z}}{\varphi(1-\varphi)} - \frac{4\tilde{h}}{(1+\varphi)^3}\right]^{-1}, \tag{5.13}$$

that is defined for $(2-\bar{z})/8\tilde{h} \leq \varphi(1-\varphi)/(1+\varphi)^3 \leq (1-\bar{z})/4\tilde{h}$. Note that for $\tilde{h} = 0$ Eq. (5.12) reduces to Eq. (2.25). The phase diagram for a pushing, zero, and pulling force $\tilde{h}$ is shown in Fig. 5.5d–f and displays, above the critical coupling strength $\tilde{J} > \tilde{J}^s_{\rm crit}$, a phase separation into a dense and dilute phase of closed bonds with compositions φ_l and φ_g, respectively. A pushing force $\tilde{h} < 0$ lifts the critical coupling and "tilts" the phase diagram towards higher density, i.e. at a given coupling $\tilde{J} > \tilde{J}^s_{\rm crit}$ the density of both phases increases. Conversely, a pulling force $\tilde{h} > 0$ lowers the critical coupling and "tilts" the phase diagram towards lower density, i.e. at a given coupling $\tilde{J} > \tilde{J}^s_{\rm crit}$ the density of both phases decreases. The biological implications of these results will be discussed in Sect. 5.5. The binodal and spinodal within the MF approximation are given in Appendix C of [1] (see Eq. (C3)–(C4) therein).

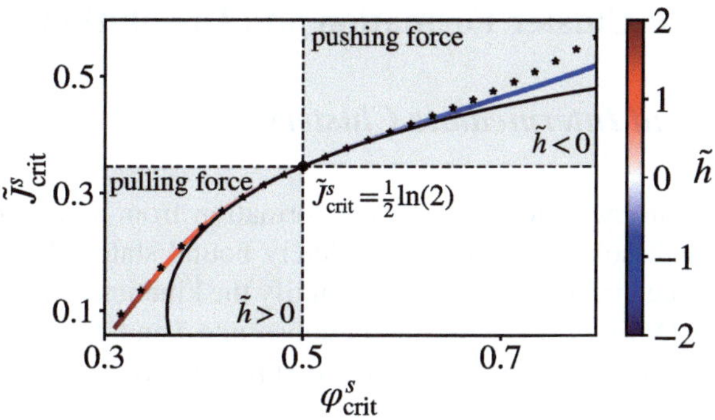

Fig. 5.6 **Statical critical point** as a function of the force $\tilde{h}$. Symbols depict the exact solution using a converged Newton's series and the gradient line depicts the quadratic approximation given by Eq. (5.29), which is very accurate for any pulling- and up to a moderate pushing force, i.e. $\tilde{h} \geq -1$. Explicit results are given in Appendix 4. The black line corresponds to the result from second order perturbation theory given by Eqs. (5.14)–(5.15) that is valid for small forces

We now address in detail the behavior of the statical critical point $(\varphi^s_{\rm crit}, \tilde{J}^s_{\rm crit})$—the point where the binodal and spinodal merge, $\tilde{J}_b(\varphi^s_{\rm crit}, \tilde{h}) = \tilde{J}_s(\varphi^s_{\rm crit}, \tilde{h}) \equiv \tilde{J}^s_{\rm crit}(\varphi^s_{\rm crit}, \tilde{h})$. The critical point denotes the onset of phase separation and is the solution of $\tilde{\mathrm{f}}'''(\varphi) = 0$, which in absence of the force yields $(\varphi^{s,0}_{\rm crit}, \tilde{J}^{s,0}_{\rm crit}) \equiv \frac{1}{2}(1, \ln \frac{\bar{z}}{\bar{z}-2})$ (see Sect. 2.8.3). In the presence of a force $\tilde{h} \neq 0$ we obtain the exact solution using a Newton's series approach [70–72] (for details regarding the Newton series see Appendix 5). The analytical result is non-trivial and is given explicitly in by Eq. (5.29) within the quadratic approximation. For small forces $|\tilde{h}| \ll 1$ we in addition derive a second order perturbation expansion $\tilde{J}^s_{\rm crit} = \tilde{J}^{s,0}_{\rm crit} - \delta\tilde{J}^s_{\rm crit}(\tilde{h}) + \mathcal{O}(\tilde{h}^3)$ and $\varphi^s_{\rm crit} = \varphi^{s,0}_{\rm crit} - \delta\varphi_{\rm crit}(\tilde{h}) + \mathcal{O}(\tilde{h}^3)$, with

$$\delta\tilde{J}^s_{\rm crit}(\tilde{h}) = \frac{8}{27}\frac{1}{\bar{z}-2}\left(\tilde{h} + \frac{2}{27}\frac{\bar{z}+2}{\bar{z}-1}\tilde{h}^2\right), \tag{5.14}$$

$$\delta\varphi_{\rm crit}(\tilde{h}) = \frac{2}{3}\frac{(\bar{z}/3)^2}{(\bar{z}-2)(\bar{z}-1)}\left[\tilde{h} + \frac{16}{9}\frac{(\bar{z}/3)^2 - \bar{z} + 1}{(\bar{z}-2)(\bar{z}-1)}\tilde{h}^2\right]. \tag{5.15}$$

The dependence of the statical critical point on the external force is depicted in Fig. 5.6. A pulling force (red) pulls the critical point towards lower $\tilde{J}$ and lower φ, whereas a pushing force (blue) effects the opposite and shifts the critical point towards larger coupling $\tilde{J}$ and higher density φ. The MF statical critical point can be derived exactly as a function of the force $\tilde{h}$, and the result is given in Appendix C of [1] (see Eq. (C13) therein).

5.4 Kinetics of Cluster Formation and Dissolution

5.4.1 Small and Intermediate Clusters

We are interested in the kinetics of cluster formation from a completely unbound state, and cluster dissolution from a completely bound state. More general initial conditions are treated in Appendix 6. We quantify the kinetics by means of the mean first passage time (MFPT) $\langle\tau_{d,f}\rangle$, where the subscripts d and f stand for **dissolution** and **formation**, respectively, and $\tau_{d,f}$ is the first passage time (FPT) defined as

$$\begin{aligned}\tau_d &\equiv \inf_t \left[\varphi(\{\sigma_i\}_t) = 0 | \varphi(\{\sigma_i\}_0) = 1\right],\\ \tau_f &\equiv \inf_t \left[\varphi(\{\sigma_i\}_t) = 1 | \varphi(\{\sigma_i\}_0) = 0\right],\end{aligned} \tag{5.16}$$

where $\{\sigma_i\}_t$ denotes the instantaneous state at time t. A cluster with N adhesion bonds has 2^N possible states $\{\sigma_i\}$. We enumerate them such that the first state corresponds to all bonds closed and the final state to all bonds broken. The **transition matrix** of the Markov chain, describing mixed Glauber-Kawasaki, has dimension $2^N \times 2^N$, whereby we must impose absorbing boundary conditions on the fully dissolved and fully bound states, respectively. An exact algebraic result for $\langle\tau_{d,f}\rangle$ is given in Eq. (5.40) in Appendix 6.1 but requires the inversion of a $(2^N - 1) \times (2^N - 1)$ sparse matrix, followed by a sum over $2^N - 1$ terms, which is feasible only for $N \lesssim 5 \times 5$.

As a result of the non-systematic cluster formation and dissolution at zero coupling $\tilde{J} = 0$, and motivated by the intuitive idea that the dynamics is dominated by low energy (i.e. minimum action) paths at large coupling $\tilde{J} \gg 1$, we make the so-called local equilibrium approximation (LEQ) to treat large clusters. Thereby we map the dynamics of the $2^N \times 2^N$ state-space onto a one-dimensional **birth-death process** for the instantaneous number of closed bonds k (see Fig. 5.7a) with **effective transition rates**

$$\bar{w}_{k\to k\pm 1} \equiv (\hat{\mathcal{Z}}_k / p_k)^{-1} \sum_{\{\sigma_i\}} \mathrm{e}^{-\beta\mathcal{H}(\{\sigma_i\})} w^{\pm}_{\text{exit}}(\{\sigma_i\}) \delta_{N_c(\{\sigma_i\}),k}, \tag{5.17}$$

where we recall that p_k represents the Glauber attempt probability in state k, and we have introduced the exit rates from configuration $\{\sigma_i\}$ in the "+" (i.e. $N_c(\{\sigma_i\}'_j) = N_c(\{\sigma_i\}) + 1$) and "−" (i.e. $N_c(\{\sigma_i\}'_j) = N_c(\{\sigma_i\}) - 1$) direction, respectively, given by

$$w^{\pm}_{\text{exit}}(\{\sigma_i\}) \equiv \sum_{j=1}^{N} w_j(\{\sigma_i\}) \delta_{N_c(\{\sigma_i\}'_j),k\pm 1}. \tag{5.18}$$

Note that only the Glauber transitions, given by Eq. (5.3), enter in Eq. (5.18). The Kawasaki transitions given by Eq. (5.5), which conserve the total number of closed bonds, enter the dynamics through the diagonal of the transition matrix as the waiting rates $\bar{w}_{k\to k} = 1 - \bar{w}_{k\to k+1} - \bar{w}_{k\to k-1}$, where the right hand side follows from conservation of probability. Within the LEQ approximation the MFPT for cluster dissolution and formation becomes

$$
\begin{aligned}
\langle \tau_d \rangle \approx \langle \tau_d^{\mathrm{le}} \rangle &= \sum_{k=0}^{N-1} \frac{1}{\bar{w}_{k\to k+1}} \sum_{l=k+1}^{N} \frac{\hat{\mathcal{Z}}_l}{\hat{\mathcal{Z}}_k} \frac{p_k}{p_l}, \\
\langle \tau_f \rangle \approx \langle \tau_f^{\mathrm{le}} \rangle &= \sum_{k=1}^{N} \frac{1}{\bar{w}_{k\to k-1}} \sum_{l=0}^{k-1} \frac{\hat{\mathcal{Z}}_l}{\hat{\mathcal{Z}}_k} \frac{p_k}{p_l},
\end{aligned}
\tag{5.19}
$$

where one can further use the DB relation $(\hat{\mathcal{Z}}_k/p_k)\bar{w}_{k\to k\pm 1} = (\hat{\mathcal{Z}}_{k\pm 1}/p_{k\pm 1})\bar{w}_{k\pm 1\to k}$ (which we prove in Appendix 6.2) to interchange the backward and forward rate in the second line and change the summation according to $\sum_{k=1}^{N} \bar{w}_{k\to k-1}^{-1} \sum_{l=0}^{k-1} (\hat{\mathcal{Z}}_l/\hat{\mathcal{Z}}_k)(p_k/p_l) \to \sum_{k=0}^{N-1} \bar{w}_{k\to k+1}^{-1} \sum_{l=0}^{k} (\hat{\mathcal{Z}}_l/\hat{\mathcal{Z}}_k)(p_k/p_l)$. In Appendix 6.3 we prove that Eq. (5.19) holds for any birth-death process where the transition rates obey DB. A comparison of the exact result given by Eq. (5.40) with the LEQ approximation in Eq. (5.19) shown in Fig. 5.7b, c demonstrates the remarkable accuracy of the approximation already for $N \sim 20$ bonds, which increases further for larger N. The reason for the high accuracy can be found in the large entropic barrier to align bonds in an unbound/bound state, effecting a local equilibration prior to complete formation/dissolution. Moreover, the LEQ approximation is expected to become asymptotically exact even for small clusters in the ideal, non-interacting limit $\tilde{J} \to 0$ as well as for $\tilde{J} \to \infty$ that is dominated by the minimum-action, "instanton" path. A further discussion of the LEQ approximation and an approximate closed form expression for Eq. (5.19) for larger systems is given in Appendix 6.

The MFPT for cluster dissolution/formation shown in Fig. 5.7b, c both display a strong and non-monotonic dependence on the coupling parameter $\tilde{J}$ with a pronounced minimum, hinting at some form of critical dynamics. As we prove below this minimum in the thermodynamic limit indeed corresponds to a **dynamical critical coupling**.

5.4.2 *Thermodynamic Limit*

We now consider dissolution and formation kinetics in very large clusters, i.e. in the limit $N \to \infty$. Note that while the MFPT formally diverges, i.e. $\lim_{N\to\infty} \langle \tau_{d,f} \rangle = \infty$, it is expected to do so in a "mathematically nice", well-defined "bulk scaling". In anticipation of an exponential scaling of relevant time-scales with the system size N, we define the **mean formation/dissolution time per bond** in the thermodynamic limit as $\langle \mathrm{t_{d,f}} \rangle \equiv \lim_{N\to\infty} \langle \tau_{d,f} \rangle^{1/N}$. Using the LEQ approximation for the MFPT given by Eq. (5.19), and assuming that the Glauber attempt probabilities p_k are strictly sub-exponential in N, we prove via a *squeezing theorem* in Appendix 6 that the exact mean dissolution and formation time per bond in the thermodynamic limit reads

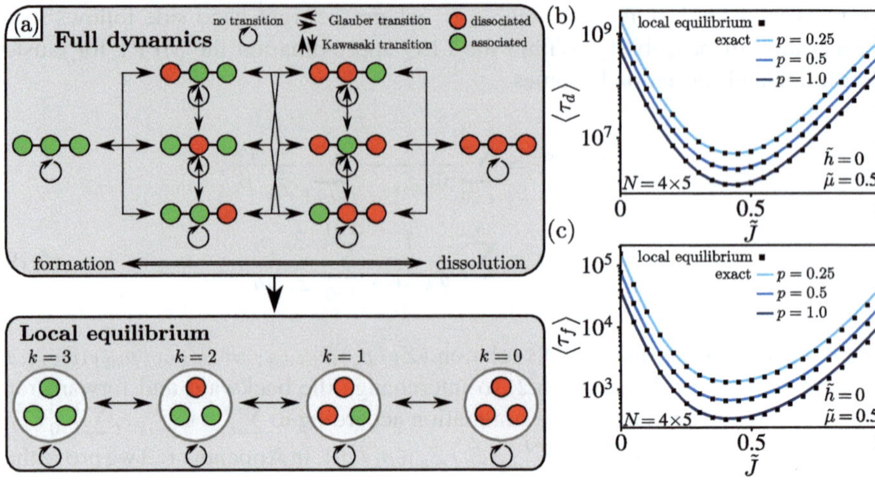

Fig. 5.7 Local equilibrium approximation and kinetics of dissolution and formation of small clusters. **a** Mapping the full dynamics onto a birth-death process. For convenience, and without any loss of generality, we here show an example for 3 adhesion bonds on a 1-dimensional lattice. The mapping holds for any lattice geometry. In the full dynamics each lattice configuration represents a different node, comprising a $2^N \times 2^N$ transition matrix, whereas in the LEQ approximation we only distinguish between states with a different number of closed/open bonds, comprising a $(N+1)\times(N+1)$ transition matrix. **b**, **c** MFPT for cluster dissolution (**b**) and formation (**c**) as a function of the coupling $\tilde{J}$ for $N = 4 \times 5$ adhesion bonds with intrinsic affinity $\tilde{\mu} = 0.5$ in the absence of a force (for nonzero force values see Fig. 5.14). Colored lines correspond to exact results obtained from Eq. (5.40) for various values of the Glauber attempt probability p, which we set to be constant $p_k \to p$, and symbols denote the LEQ approximation Eq. (5.19) evaluated with the exact $\hat{\mathcal{Z}}_k$ and $\bar{w}_{k\to k\pm 1}$ obtained from brute-force summation of Eqs. (5.6) and (5.17)

$$\langle \mathrm{t_{d,f}} \rangle = \mathrm{e}^{\tilde{\mathrm{f}}(\varphi^{d,f}_{\max})-\tilde{\mathrm{f}}(\varphi^{d,f}_{\min})} \equiv \mathrm{e}^{\Delta\tilde{\mathrm{f}}^{\dagger}}, \tag{5.20}$$

where

$$\begin{aligned}\varphi^{d}_{\max} &\equiv \sup_{\varphi<1}\tilde{\mathrm{f}}(\varphi), \quad \varphi^{d}_{\min} \equiv \inf_{\varphi>\varphi^{d}_{\max}}\tilde{\mathrm{f}}(\varphi),\\ \varphi^{f}_{\max} &\equiv \sup_{\varphi>0}\tilde{\mathrm{f}}(\varphi), \quad \varphi^{f}_{\min} \equiv \inf_{\varphi<\varphi^{f}_{\max}}\tilde{\mathrm{f}}(\varphi).\end{aligned} \tag{5.21}$$

Equation (5.20) shows that the mean first passage per bond in the thermodynamic limit is determined exactly by the largest left/right-approaching free energy barrier between the initial and final point, and is completely independent of the Glauber attempt probability p_k. We obtain analytical results for Eqs. (5.20) and (5.21) for arbitrary $\tilde{J}$, $\tilde{\mu}$ and $\tilde{h}$. Since these results are somewhat complicated for $\tilde{\mu} > 0$ and $\tilde{h} \neq 0$ we present them in Appendix 6 and Fig. 5.15. In the force-free case with zero intrinsic affinity, i.e. $\tilde{\mu} = \tilde{h} = 0$, they turn out to be surprisingly compact and given by

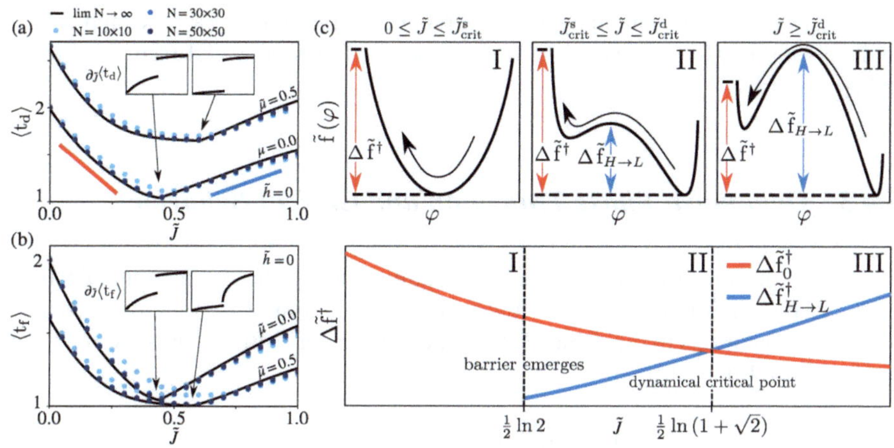

Fig. 5.8 Master scaling of mean dissolution and formation times per bond for finite clusters and in the thermodynamic limit, and the origin of the dynamical critical point. $\langle t_{d,f} \rangle$ for cluster dissolution (**a**) and formation (**b**) as a function of the coupling $\tilde{J}$ for a pair of intrinsic affinities $\tilde{\mu} = 0$ and $\tilde{\mu} = 0.5$ and various cluster sizes (symbols) as well as the thermodynamic limit (lines) in the absence of an external force; Symbols are evaluated with the LEQ approximation Eqs. (5.19) using $\hat{\mathcal{Z}}_k^{BG}$ [Eqs. (5.6) and (5.8)] and $\bar{w}_{k\to k+1}$ from Eq. (5.17). The discrepancy between the lines and symbols is due to finite-size effects. **c** In the thermodynamic limit the mean dissolution/formation time $\langle t_{d,f} \rangle$ depends only on the *largest* free energy barrier (see Eq. (5.20)). For small coupling (regime I) the latter corresponds to the difference between the free energy minimum and the fully dissolved or bound configuration, $\Delta\tilde{f}^\dagger = \Delta\tilde{f}^\dagger_{0,1}$, respectively. At the statical critical coupling value, J^s_{crit}, (onset of regime II) a free energy barrier emerges separating the meta-stable from the stable phase, $\Delta\tilde{f}^\dagger_{H\rightleftarrows L}$, but the largest free energy barrier is still $\Delta\tilde{f}^\dagger = \Delta\tilde{f}^\dagger_{0,1}$. At the *dynamical critical coupling*, $\tilde{J}^d_{crit}$, (onset of regime III) the free energy barrier separating the meta-stable from the stable phase becomes dominant, $\Delta\tilde{f}^\dagger = \Delta\tilde{f}^\dagger_{H\rightleftarrows L}$. The depicted free energy landscapes $\tilde{f}(\varphi)$ correspond to Eq. (5.10) with $\tilde{\mu} = 0.05$ and $\tilde{h} = 0$

$$
\langle t_{d,f} \rangle = \begin{cases} 2e^{-2\tilde{J}}\cosh^2\tilde{J}, & 0 \le \tilde{J} \le \frac{1}{2}\ln 2 \\ 4\dfrac{\sinh^2 2\tilde{J}}{e^{4\tilde{J}}-2}, & \frac{1}{2}\ln 2 \le \tilde{J} \le \frac{1}{2}\ln(1+\sqrt{2}) \\ 8e^{2\tilde{J}}\dfrac{\sinh^2\tilde{J}}{e^{4\tilde{J}}-2}, & \tilde{J} \ge \frac{1}{2}\ln(1+\sqrt{2}), \end{cases} \tag{5.22}
$$

such that for $\tilde{J} = 0$ and $\tilde{J} \to \infty$ we have $\langle t_{d,f} \rangle = 2$ being the maximum, and the minimum occurs at $\tilde{J} = \ln(1+\sqrt{2})/2$ where $\langle t_{d,f} \rangle = (4/7)(2\sqrt{2}-1)$. Figure 5.8a,b shows a comparison of the prediction of Eq. (5.20) with the results for finite systems given by Eq. (5.19) rescaled according to $\langle \tau_{d,f} \rangle^{1/N}$. Already for $N = 900$ a nearly complete collapse to the thermodynamic limit (5.20) is observed for both, cluster formation as well as dissolution. The MF analogue of Eq. (5.22) is given by Eq. (E35) in [1] for general $\bar{z}$ and remarkably has a universal minimum (i.e. $\bar{z}$-independent) value of $\langle t_{d,f} \rangle_{MF} \approx 1.0785$ at the dynamical critical coupling $\tilde{J} = 2\ln(2)/\bar{z}$. Moreover, $\langle \mathrm{t_{d,f}} \rangle_{\mathrm{MF}}$ displays an unphysical divergence in the limit $\tilde{J} \to \infty$ which relates to the diverging MF free energy barrier as discussed in Sect. 2.11.1 (see also Fig. 18 in [1]).

5.4.3 Dynamical Phase Transition and Critical Behavior

Strikingly, the mean dissolution and formation time in the thermodynamic limit (5.20) display a discontinuity as a function of the coupling $\tilde{J}$ (see jumps in $\partial_{\tilde{J}} \langle \mathrm{t_{d,f}} \rangle$ depicted in the insets in Fig. 5.8a, b). In particular, for zero affinity and external force we find from Eq. (5.22)

$$\lim_{\tilde{J} \nearrow \frac{1}{2}\ln(1+\sqrt{2})} \partial_{\tilde{J}} \langle \mathrm{t_{d,f}} \rangle = -(4/7)^2(13\sqrt{2} - 17)$$

$$\lim_{\tilde{J} \searrow \frac{1}{2}\ln(1+\sqrt{2})} \partial_{\tilde{J}} \langle \mathrm{t_{d,f}} \rangle = (8/7^2)(9\sqrt{2} - 8).$$

This implies the existence of a first order dynamical phase transition at the dynamical critical coupling $\tilde{J}^d_{\mathrm{crit}}$ and hence a qualitative change in the dominant dissolution/formation pathway. Coincidentally, the BG *dynamical critical point* for $\tilde{\mu} = \tilde{h} = 0$ coincides with the exact (Onsager's) statical critical point for the two-dimensional zero-field Ising model [73] (for a more detailed discussion see Appendix 6). Similarly, the MF dynamical critical point for $\tilde{\mu} = \tilde{h} = 0$ coincides with the BG statical critical point as shown in [1]. Strikingly, the dynamic critical point always corresponds to the minimum of $\langle \mathrm{t_{d,f}} \rangle$. The explanation of the physics underneath the dynamical phase transition and the meaning of $\tilde{J}^d_{\mathrm{crit}}$ is given in Fig. 5.8c.

The qualitative behavior of $\langle \mathrm{t_{d,f}} \rangle$ has three distinct regimes. In regime I, where $0 \le \tilde{J} < \tilde{J}^s_{\mathrm{crit}}$, the free energy landscape $\tilde{\mathrm{f}}(\varphi)$ has a single well and according to Eq. (5.20) $\langle \mathrm{t_{d,f}} \rangle$ is determined by $\Delta\tilde{\mathrm{f}}^{\dagger}_{0,1}$—the free energy difference between the minimum and the absorbing point (i.e. $\varphi = 0$ for dissolution and $\varphi = 1$ for formation, respectively). $\Delta\tilde{\mathrm{f}}^{\dagger}_{0,1}$ is a decreasing function of $\tilde{J}$.

At the statical critical coupling $\tilde{J}^s_{\mathrm{crit}}$, which marks the onset of regime II, a second free energy barrier emerges delimiting the phase-separated low (L) and a high (H) density phase. We denote this free energy barrier by $\Delta\tilde{\mathrm{f}}^{\dagger}_{H \rightleftarrows L}$ where $\to$ and $\leftarrow$ stand for dissolution and formation, respectively. $\Delta\tilde{\mathrm{f}}^{\dagger}_{H \rightleftarrows L}$ is an increasing function of $\tilde{J}$. In regime II, that is when $\tilde{J}^s_{\mathrm{crit}} \le \tilde{J} < \tilde{J}^d_{\mathrm{crit}}$, the dissolution and formation first evolve through a (thermodynamic) phase transition and, finally, must also surmount the second, predominantly entropic barrier to the complete dissolved/bound state. In regime II, as in regime I, the largest free energy barrier remains the free energy difference between the minimum and the absorbing point, i.e. $\Delta\tilde{\mathrm{f}}^{\dagger}_{0,1} > \Delta\tilde{\mathrm{f}}^{\dagger}_{H \rightleftarrows L}$.

Exactly at the *dynamical critical coupling* $\tilde{J}^d_{\mathrm{crit}}$ the two barriers become identical, $\Delta\tilde{\mathrm{f}}^{\dagger}_{0,1} = \Delta\tilde{\mathrm{f}}^{\dagger}_{H \rightleftarrows L}$ and for $\tilde{J} > \tilde{J}^d_{\mathrm{crit}}$ we always have $\Delta\tilde{\mathrm{f}}^{\dagger}_{0,1} < \Delta\tilde{\mathrm{f}}^{\dagger}_{H \rightleftarrows L}$. Therefore, in regime III the rate-limiting event becomes the phase transition itself, whereas

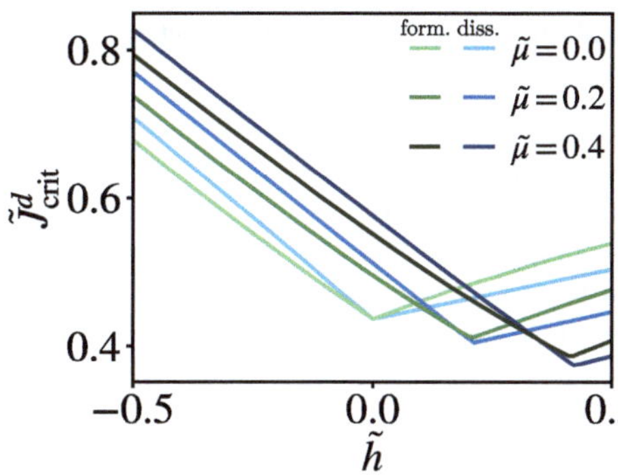

Fig. 5.9 Dynamical critical coupling $\tilde{J}^d_{\rm crit}$ as a function of the external force $\tilde{h}$ for several values of the intrinsic binding-affinity $\tilde{\mu}$. Note that $\tilde{J}^d_{\rm crit}$ is non-monotonic with a global minimum whose location depends on $\tilde{\mu}$ and $\tilde{h}$

the fully dissolved/bound state is thereupon reached by typical density fluctuations. Since $\Delta\tilde{\rm f}^\dagger_{0,1}$ decreases with $\tilde{J}$ while $\Delta\tilde{\rm f}^\dagger_{H\rightleftarrows L}$ increases with $\tilde{J}$, the mean dissolution/formation time per bond at the dynamical critical coupling $\tilde{J}^d_{\rm crit}$ must be minimal. This explains the dynamical phase transition completely.

Note that the dynamical phase transition is preserved under equilibrium initial conditions, i.e. upon considering $\varphi(\{\sigma_i\}_0) = \varphi^0_{L,H}$ in Eq. (5.16), where $\varphi^0_{L,H}$ is the (meta)-stable minimum in the high or low density region for respectively cluster dissolution or formation. In the thermodynamic limit the equilibration time from the initial condition $\varphi(\{\sigma_i\}) = 0 \vee 1$ to the (meta)-stable minimum $\varphi^0_{L,H}$ becomes exponentially suppressed w.r.t. the total transition time, which renders $\langle {\rm t_{d,f}} \rangle$ unchanged.

The dependence of $\tilde{J}^d_{\rm crit}$ on $\tilde{\mu}$ and $\tilde{h}$ is determined in the form of a Newton's series in Appendix 6, and is depicted in Fig. 5.9. Depending on the intrinsic affinity $\tilde{\mu}$, the dependence of $\tilde{J}^d_{\rm crit}$ may be non-monotonic. Note that in contrast to the statical critical coupling $\tilde{J}^{\rm s}_{\rm crit}$ which is independent of $\tilde{\mu}$, the dynamical critical coupling $\tilde{J}^{\rm d}_{\rm crit}$ depends on the particular value of $\tilde{\mu}$.

5.5 Mechanical Regulation of Cell Adhesion

Our results tie the effective bending rigidity, κ, and in turn interactions between neighboring adhesion bonds, $\tilde{J} \propto \kappa^{-1/2}$ (see Appendix 2), to the collective phase behavior of adhesion clusters at equilibrium, and to distinct dynamical phases of cluster dissolution and formation. Based on the quantitative relationship between the coupling strength $\tilde{J}$ and bending ridigity κ given by Eq. (5.25), and an order-of-magnitude estimation of the relevant parameters listed in Table 5.1, we find that the coupling strength in cellular systems lies within the range $0 \lesssim \tilde{J} \lesssim 2.5$. Notably, both the statical and dynamical critical point at moderate values of the external force and/or intrinsic binding-affinity lie within said range (see Figs. 5.6 and 5.9). Yet, it remains to be explained why a near-critical coupling may be beneficial for cells, and how it may be regulated.

Table 5.1 Estimated parameter to determine the range of $\tilde{J}$ through Eq. (5.25)

Parameter	Symbol	Estimated value/range	Sources
Spring constant	βk	$\sim 10^{-2}$ (nm^{-2})	[39, 82, 83]
Non-specific interaction strength	$\beta\gamma$	$\sim 10^{-5}$ (nm^{-4})	[39]
Bond separation distance	$h_0 - l_0$	25–50 (nm)	[82, 84, 85]
Bending rigidity	$\beta\kappa$	4–400	[50, 79, 80, 86]

Our results provoke the hypothesis that the membrane rigidity (and hence the coupling strength) may lie close to the statical critical value for quasi-static, and near the dynamical critical value for transient processes. Mechanical regulation of the bending rigidity can be achieved through hypotonic swelling [74], (de)polymerization of the F-actin network [75, 76], by decoupling the F-actin network from the plasma membrane [50], through changes of the membrane composition [77–80] or integral membrane proteins [78], membrane-protein activity [81], temperature modulation [18, 26, 79], and acidosis [37], to name but a few. Moreover, it has been shown experimentally that temperature modulations affects adhesion strength through changes in membrane fluidity [18], cell elasticity [26], or via a temperature cooperative process [16], albeit the denaturation of the binding proteins also provides a possible explanation [27].

Below we argue that the *change* in the response of a cell to a perturbation, defined as a change in the equilibrium binding strength or association/dissociation rates, is largest near criticality. This results in either a very small or very large response, depending on the change of the underlying parameter. Here we follow the same kind of reasoning as rooted in the criticality hypothesis, which states that systems undergoing an order–disorder phase transition achieve the highest trade-off between robustness an flexibility around criticality [87].

5.5.1 Criticality at Equilibrium

In Fig. 5.10a we depict how oscillations in the coupling strength (arising through oscillations in the bending rigidity κ) around the statical critical point affect the average fraction of closed bonds. Similar oscillatory patterns and their effect on the adhesion strength have been observed in vascular smooth muscle cells, where changes in the bending rigidity were concerted by the remodeling of the actin cytoskeleton [25, 52, 53]. Minute changes in the amplitude, $\delta\tilde{J}$, can drive the systems's behavior from oscillations within a dense phase with $\langle\varphi(t)\rangle > 0.5$ to intermittent periods of nearly complete dissolution (compare full and dashed lines in Fig. 5.10a). Hence we

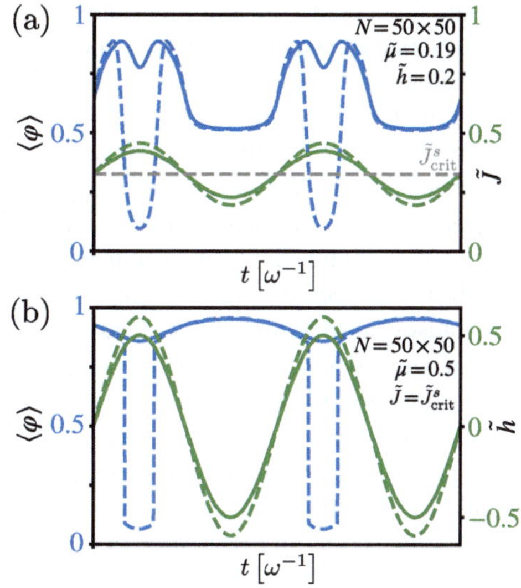

Fig. 5.10 Equilibrium response to changes in cell stiffness and external force. Equilibrium response of the average fraction of closed bonds $\langle\varphi(t)\rangle$ to a slow (quasi-static), periodic modulation of **a** the coupling $\tilde{J}(t)=\tilde{J}+\delta\tilde{J}\sin(\omega t)$ (and hence membrane stiffness) and **b** the external force $\tilde{h}(t)=\tilde{h}+\delta\tilde{h}\sin(\omega t)$. Note the strong sensitivity of the response near the *statical* critical coupling $\tilde{J}^s_{\mathrm{crit}}$

find that the response (i.e. $\langle\varphi(t)\rangle$) is most sensitive to a change in the amplitude $\delta\tilde{J}$ when $\tilde{J}$ lies close to the statical critical point.

Similarly, in Fig. 5.10b we show the response of $\langle\varphi(t)\rangle$ to a mechanical perturbation oscillating quasi-statically between a pulling and a pushing force, $\tilde{h}(t)=\tilde{h}+\delta\tilde{h}\sin(\omega t)$ (for practical examples see e.g. [88, 89]). Such mechanical perturbations can for example arise through changes in active stresses generated within the cytoskeleton [90]. Here as well, a small change in the force $\delta\tilde{h}$ acting on the cluster, can lead to stark differences in the cluster stability $\langle\varphi(t)\rangle$. The sensitivity to a change in the force is most amplified near the statical critical coupling $\tilde{J}^s_{\mathrm{crit}}$ (compare full and dashed lines in Fig. 5.10b), where a small change in the amplitude, $\delta\tilde{h}$, can cause intermittent periods of essentially complete cluster detachment.

Drastic changes in the average number of closed bonds have been observed experimentally in adhesion frequency assays and single-molecule microscopy [11, 58]. There it was shown that binding affinities and binding dynamics for a T-cell receptor (TCR) interacting with the peptide-major histocompatbility complex (pMHC) are more than an order of magnitude smaller in solution (i.e. in 3D) as compared to when they are anchored to a cell membrane (i.e. in 2D). One possible contribution to the discrepancy between the 3D and 2D binding kinetics is the difference in the reduction of the entropy upon binding, which is larger in 3D than in 2D [31]. However, it has been explicitly remarked that this contribution alone does *not* explain the measured difference in the binding affinities [31]. The authors of Ref. [11, 58] rationalize these differences in binding in terms of a cooperativity between neighboring TCRs due to the anchoring membrane. In particular, Fig. 5.11a in the SM of Ref. [58] shows the adhesion frequency $P_a(t_c)\in[0,1]$, defined as the fraction of observed adhesion events between the TCR and pMHC as a function of the contact time t_c between the anchoring membranes, derived from MC simulations. Upon introducing a heuristic neighbor-dependent amplification factor in the binding rates, the authors observe an amplification of the adhesion frequency P_a (compare squares with diamonds), indicating an increase in binding events in agreement with their experimental observations.

We relate our results to the observations in Ref. [58] by recalling the relation between P_a and $\langle N_c\rangle = N\langle\varphi\rangle$, i.e. $P_a^{\mathrm{ss}} \equiv \lim_{t_c\to\infty} P_a(t_c) = 1 - \exp(-N\langle\varphi\rangle)$ (see [91] as well as Eqs. (1) and (2) in [58]). In our model the aforementioned amplification factor arises naturally from a nonzero coupling strength $\tilde{J}$ due to the anchoring membrane. Indeed, in Fig. 5.4e an increase in $\tilde{J}$ leads to an increase in $\langle\varphi\rangle$, which in turn causes an increase of the steady state adhesion frequency P_a^{ss}. Hence, we find that the amplification factor in [58] and coupling $\tilde{J}$ in our model have the same effect on the adhesion frequency.

A similar observation was made in [30] on the basis of a detailed analysis of the binding affinities of the adhesion receptor CD16b placed in three distinct environments: red blood cells (RBCs), detached Chinese hamster ovary (CHO) cells, and K562 cells. Based on Fig. 5.4a, b in [30] the adhesion frequency for RBCs is around a 15-fold larger than for CHO and K562 cells. In the discussion the authors point towards the modulation of surface smoothness as an explanation for the observed differences in adhesion frequency [30]. Since K562 cells are known to have a larger bending rigidity than RBCs [92, 93] (we were unfortunately not able to find the corresponding information for CHO cells in the existing literature), it is expected that the coupling strength $\tilde{J}$ is generally higher in the latter (see Appendix 2), which provides a potential explanation for the observed difference in adhesion frequencies between RBCs and K562 cells.

5.5.2 *Criticality in Kinetics*

Many biological processes [94–97] and experiments [98–100] involve adhesion under transient, non-equilibrium conditions, where cells can detach completely from a substrate (for a particular realization with a constant force see [98]). The duration of these transients may be quantified by the mean dissolution and formation time, $\langle t_{d,f}\rangle$ (see Fig. 5.8). Imagine that the cell can change the bending rigidity by an amount $\Delta\kappa$ that in turn translates into a change in coupling, $\tilde{J}' = \tilde{J} + \Delta\tilde{J} \propto 1/\sqrt{\kappa + \Delta\kappa}$. If the mechanical regulation is to be efficient, a small change of $\Delta\tilde{J}$ should effect a large change of $\langle t_{d,f}\rangle$.

The efficiency of the regulation, expressed as the change of mean dissolution/formation time in response to a change $\Delta\tilde{J}$, $\Delta\langle t_{d,f}\rangle \equiv \langle t_{d,f}(\tilde{J}+\Delta\tilde{J})\rangle - \langle t_{d,f}(\tilde{J})\rangle$, is shown in Fig. 5.11. The results demonstrate that the regulation is most efficient, that is gives the largest change, when $\tilde{J}$ is poised near the *dynamical* critical coupling, $\tilde{J} \simeq \tilde{J}^d_{\mathrm{crit}}$, regardless of the magnitude of the change $\Delta\tilde{J}$. Recall that the formation and dissolution rate, $1/\langle t_f\rangle$ and $1/\langle t_d\rangle$, respectively, are highest at the dynamical critical coupling (see Fig. 5.8). Therefore, we not only find the largest response to a change in $\tilde{J}$, but also the fastest formation and dissolution kinetics at the dynamical critical coupling.

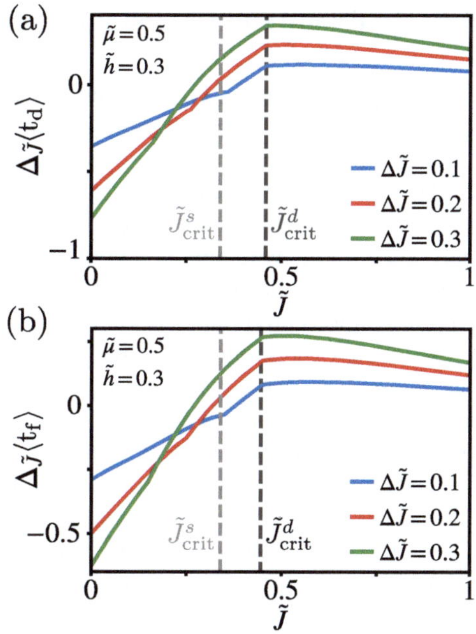

Fig. 5.11 Kinetic response to changes in cell stiffness. Change in mean dissolution (**a**) and formation (**b**) time per bond, $\Delta_{\tilde{J}}\langle t_{d,f}\rangle \equiv \langle t_{d,f}(\tilde{J}+\Delta\tilde{J})\rangle - \langle t_{d,f}(\tilde{J})\rangle$, as a response to a change $\Delta\tilde{J}$ of the coupling, as a function of $\tilde{J}$ for various $\Delta\tilde{J}$. The kinetic response is largest for $\tilde{J}$ near the *dynamical critical coupling*

An example where fast kinetic (un)binding and a large sensitivity to the bending rigidity can be beneficial is found in tumor cells that undergo metastasis—the process through which tumor cells spread to secondary locations in the host's body. Recent studies suggest that cancer cells are mechanically more compliant than normal, healthy cells [101]. Moreover, experiments with magnetic-tweezers have shown that membrane stiffness of patient tumor cells and cancer cell-lines inversely correlates with their migration and invasion potential [55], and an increase of membrane rigidity alone is sufficient to inhibit invasiveness of cancer cells [86]. Cells with the highest invasive capacity were found to be five times less stiff than cells with the lowest migration and invasion potential, but the underlying mechanism behind this correlation remained elusive [55].

Based on our results a decrease in the bending rigidity, and hence the membrane stiffness, can alter both, the equilibrium strength of adhesion (see Figs. 5.4 and 5.10) as well as the kinetics of formation and dissolution of adhesion domains (see Figs. 5.8 and 5.11). This may provide a clue about the mechanical dysregulation of cell adhesion in metastasis in terms of a softening of the cell membrane.

5.6 Criticality in the Ising Model

By setting $\tilde{h} = 0$ in Eq. (5.1) our findings, and in particular the uncovered dynamical phase transition, also provide new insight into equilibrium and kinetic properties of the Ising model in the presence of a uniform external magnetic field.

The equilibrium properties of the two-dimensional Ising model in the absence of a magnetic field, such as the total free energy per spin, statical critical point, and binodal line were obtained in the seminal work by Onsager [73]. The effect of a uniform magnetic field has mostly been studied numerically [102, 103], e.g. by MC simulations [104] and renormalization group theory [105], but hitherto no exact closed-form expression for the free energy per spin has been found. On the BG level the free energy density, binodal line, spinodal line, and statical critical point were known [106], but to our knowledge we are the first to provide an exact closed-form expression for the EOS in the presence of a uniform magnetic field (see Appendix 5).

The kinetics of the two-dimensional Ising model have been studied in the context of magnetization-reversal times (i.e. the time required to reverse the magnetization) [107–109], nucleation times [110, 111], and critical slowing down [112, 113]. Here we report a new type of dynamical critical phenomenon related to a first-order discontinuity and a global minimum of the magnetization reversal time at the concurrent dynamical critical point (see Fig. 5.8), which is fundamentally different from the statical critical point. The dynamical phase transition reflects a qualitative change in the instanton path towards magnetization reversal, and has not been reported before.

In Table 5.2 we summarize the values of the statical and dynamical critical points obtained by the MF and BG approximation in the absence of a magnetic field, and for a general coordination number $\bar{z}$ (for a derivation of the dynamical critical points see Appendix 6 and Appendix 9 in [1]. We also state the exact statical critical point of the two-dimensional Ising model. Conversely, the exact dynamical critical point of the two-dimensional Ising model remains unknown as it requires the exact free energy density as a function of the fraction of down spins (see Eq. (5.10) for the result within the BG approximation). A lower bound on the dynamical critical point is set by the statical critical point, as the latter denotes the onset of an interior local maximum that is required for the dynamical critical point (see Fig. 5.8). The exact

Table 5.2 Statical and dynamical critical point as a function of the coordination number $\bar{z}$ obtained with the MF and BG approximation, alongside the exact statical critical point $\tilde{J}^{s}_{\text{crit}}$ for the two-dimensional Ising model at zero field and binding-affinity $\tilde{\mu} = \tilde{h} = 0$

Critical points		
Approximation	$\tilde{J}^{s}_{\text{crit}}$	$\tilde{J}^{d}_{\text{crit}}$
MF	$1/\bar{z}$	$2\ln(2)/\bar{z}$
BG	$\ln(\bar{z}/(\bar{z}-2))/2$	$-\ln(2^{1-2/\bar{z}}-1)/2$
Exact 2D	$\ln\left(1+\sqrt{2}\right)/2$	$\geq \tilde{J}^{s}_{\text{crit}}$

The exact dynamical critical point remains unknown; a lower bound is given by the Onsager statical critical point dynamical critical point may provide further insight into the nature of the dynamical phase transition. Moreover, it also sets a lower bound on the magnetization reversal times per spin in ferromagnetic systems in the absence of an external force.

5.7 Concluding Remarks

The behavior of individual [2] and non-interacting [3–5, 64] adhesion bonds under force, the effect of the elastic properties of the substrate and pre-stresses in the membrane [35, 65], as well as the physical origin of the interaction between opening and closing of individual adhesion bonds due to the coupling with the fluctuating cell membrane [36, 38–42, 114, 115], are by now theoretically well established. However, in order to understand the importance of these interactions and their manifestation for the mechanical regulation of cell adhesion in and out of equilibrium, one must go deeper, and disentangle the response of adhesion clusters of all sizes to external forces and how it becomes altered by changes in membrane stiffness. This is paramount because interactions strongly change the physical behavior of adhesion clusters under force both, qualitatively as well as quantitatively.

Founded on firm background knowledge [2–5, 35, 36, 38–42, 64, 65, 114, 115], our explicit analytical results provide deeper insight into cooperative effects in cell-adhesion dynamics and integrate them into a comprehensive physical picture of cell adhesion under force. We considered the full range of CAM binding-affinities and forces, and established the phase behavior of two-dimensional adhesion clusters at equilibrium as well as the kinetics of their formation and dissolution.

We have obtained, to the best of our knowledge, the first theoretical results on equilibrium behavior and dynamic stability of adhesion clusters in the thermodynamic limit beyond the MF-level (existing studies, even those addressing non-interacting adhesion bonds [4, 5, 64] are limited to small clusters sizes [35, 39, 40, 65]). We explained the complete thermodynamic phase behavior, including the co-existence of dense and dilute adhesion domains, and characterized in detail the corresponding critical behavior.

We demonstrated conclusively the existence of a seemingly new kind of dynamical phase transition in the kinetics of adhesion cluster formation and dissolution, which arises due to the interactions between the bonds and occurs at a critical coupling $\tilde{J}^d_{\mathrm{crit}}$, whose value depends on the external force $\tilde{h}$ and binding-affinity $\tilde{\mu}$. At the dynamical critical coupling $\tilde{J}^d_{\mathrm{crit}}$, and in turn critical bending rigidity $\kappa^d_{\mathrm{crit}} \propto (\tilde{J}^d_{\mathrm{crit}})^{-2}$, the dominant formation and dissolution pathways change qualitatively. Below $\tilde{J}^d_{\mathrm{crit}}$ the rate-determining step for cluster formation and dissolution is the surmounting of the (mostly) entropic barrier to completely bound and unbound states, respectively. Conversely, above $\tilde{J}^d_{\mathrm{crit}}$ the thermodynamic phase transition between the dense and dilute phase for dissolution, and between the dilute and dense phase for cluster formation, becomes rate-limiting, whereas the completely bound and unbound states, respectively, are thereupon reached by typical density fluctuations.

In an experimental setup we expect that the non-monotonicity between the MFPT to cluster dissolution/formation and the coupling strength $\tilde{J}$–which is a direct reflection of the dynamical critical transition–could be measured. By determining the dissolution/formation time (in the absence or presence of an external driving force) for an ensemble of cells which are adherent to a stiff interface, one could calculate the MFPT to cluster dissolution/formation. Then, upon varying the membrane rigidity through varying the membrane composition (i.e. increasing the cholesterol composition increases the membrane rigidity [79]), one could vary the coupling strength, and thereby potentially obtain the MFPT as a function of the coupling strength $\tilde{J}$.

We discussed the biological implications of our results in the context of mechanical regulation of the bending rigidity around criticality. Based on our results we have suggested that the response of a cell to a change in the bending rigidity may be largest near the statical critical point for quasi-static processes, and near the dynamical critical point for transient processes. This observation agrees with the *criticality hypothesis* and might expand the list of biological processes hypothesized to be poised at criticality [116].

Finally, we discussed the implications of our result for the two-dimensional Ising model. The observed dynamical phase transition is related to a first-order discontinuity in the magnetization reversal time, and the exact dynamical critical point for the two-dimensional Ising model remains elusive (see Table 5.2).

5.7.1 Model Limitations

We now remark on the limitations of our results. The mapping onto a lattice gas/Ising model (i.e. Eq. (5.1) and Appendix 2; see also [39]) may not apply to genuinely floppy membranes encountered in biomimetic vesicular systems [6, 7]. Moreover, since we only allow for two possible states of the bonds, i.e. associated and dissociated, we neglect any internal degrees of freedom (e.g. orientations of the bonds) which may contribute to the entropy loss upon binding [31], thereby changing the free energy.

Likewise, the assumption of an equally shared force is generally good for stiff membranes (stiffened by the presence of, or anchoring to, the stiff actin cytoskeleton [9]) or stiff membrane/substrate pairs, flexible individual bonds, low bond-densities, or the presence of pre-stresses exerted by the actin cytoskeleton [35, 64, 65]. In Appendix 3 we provide an analysis of the effect of a non-uniform force load. Based on this analysis we find that in the case of rather floppy membranes, corresponding to large values of the coupling strength $\tilde{J} = \mathcal{O}(1)$, the difference between a uniform and a non-uniform force load is negligible for a broad range of realizations of the non-uniform force distribution. Only under the extreme, non-physiological condition that the ratio of forces experienced by inner and outer bonds is larger than an order of magnitude, we observe significant differences. Therefore, the dependencies of the statical and dynamical critical points on the external force (see Figs. 5.6 and 5.9, respectively) are expected to remain valid for a non-uniform force distribution over a large range of force magnitudes.

In their present form our results may not apply to conditions when cells actively contract in response to a mechanical force on a timescale comparable to cluster assembly or dissolution [117], as well as situations in which cells actively counteract the effect of an external pulling force and make adhesion clusters grow.

Finally, throughout we have considered clusters consisting of so-called **slip-bonds**, whereas cell adhesion may also involve **catch-bonds** that dissociate slower in the presence of sufficiently large pulling forces [118]. The reason lies in a second, alternative dissociation pathway that becomes dominant at large pulling forces [119–122]. Our results therefore do not apply to focal adhesions composed of catch-bonds and would require a generalization of the Hamiltonian (5.1–5.2) and rates (5.3–5.5). These open questions are beyond the scope of the present work and will be addressed in forthcoming publications.

Appendices

Appendix 1: Data Availability

The open source code for the evaluation of the EOS and MFPT to cluster dissolution and formation for finite-size systems is available at [123].

Appendix 2: Membrane Rigidity and the Coupling Strength

Here we provide a quantitative relation between the effective bending rigidity κ and the coupling strength $\tilde{J}$ based on the results of Ref. [39]. Consider a set of adhesion bonds at fixed positions $\{\mathbf{r}_i\}$ coupled to a fluctuating membrane. The effective bending rigidity quantifies the amount of energy needed to change the membrane curvature, and is supposed to depend on the membrane composition [79, 80], state of the actin network [50], and other intrinsic factors that determine the mechanical stiffness of the cell. Let $\{b_i\}$ describe the state of all bonds, where $b_i = 1$ denotes a closed and $b_i = 0$ an open bond. The bonds are represented by springs with constant k, resting length l_0, and binding energy ϵ_b. Non-specific interactions between the membrane and the opposing substrate are described by a harmonic potential with strength γ, which arises from a Taylor expansion around the optimal interaction distance h_0 between the membrane and the substrate. Assuming a timescale separation between the opening/closing of individual bonds and membrane fluctuations, the following partition function for the state of bonds $\{b_i\}$ can be derived [39]

$$\mathcal{Z} = \sum_{\{b_i\}} \exp\left(\sum_{i\neq j} \tilde{J}_{ij} b_i b_j + \tilde{\mu} \sum_{i=1}^{N} b_i\right) + \mathcal{O}\left(\frac{k^2}{\gamma\kappa}\right), \tag{5.23}$$

where $\tilde{\mu}$ plays the role of an intrinsic binding-affinity and $\tilde{J}_{ij}$ is an effective interaction between the bonds given by

$$\tilde{J}_{ij} \equiv \frac{\beta k^2 (h_0 - l_0)^2}{16\sqrt{\gamma\kappa}} m\left(|\mathbf{r}_i - \mathbf{r}_j|\right), \tag{5.24}$$

with $m(r) = -(4/\pi)\mathrm{kei}_0\left(r(\kappa/\gamma)^{1/4}\right)$, and $\mathrm{kei}_0(\mathrm{x})$ is a Kelvin function defined as $\mathrm{kei}_0(x) \equiv \mathrm{Im}\, K_0(xe^{3\pi i/4})$ where $K_0(z)$ is the zero-order modified Bessel function of the second kind [39]. A systematic comparison of the EOS φ of the full/explicit model (i.e. reversible adhesion bonds coupled to a dynamic, fluctuating membrane) and of the lattice gas governed by Eq. (5.24) has been carried out in [39]. Using the following set of parameter values $\beta\kappa = 80$, $\beta\gamma = 10^{-5}$ nm^{-4}, $\beta k = 2.25 \times 10^{-2}$ nm^{-2}, $h_0 - l_0 = 45.9 - 50.3$ nm, and $m(1.5) = 0.42194(6)$, corresponding to a coupling strength of $\tilde{J} \approx 1.0 - 1.2$, the authors found a *quantitative* agreement between the full and lattice gas models (see Fig. 5 in Ref. [39]). Note that the lattice gas model becomes exact in the limit $\kappa \to \infty$.

Our effective Hamiltonian, given by Eq. (5.1), is directly derived from Eq. (5.23) by considering the following arguments: First we note that the effective interaction $\tilde{J}_{ij}$ decays exponentially fast as a function of the lattice distance between the bonds, and therefore it suffices to only take into account nearest neighbor interactions [39]. Moreover, since we place the adhesion bonds on a lattice with equidistant vertices, the position dependence in Eq. (5.24) drops out and we get $|\mathbf{r}_i - \mathbf{r}_j| = \Delta r$. Finally, upon introducing the variables $\sigma_i \equiv 1 - 2b_i \in [-1, 1]$, and applying the transformations $\tilde{\mu} \to \tilde{\mu} - \bar{z}\tilde{J}$ and $\tilde{J} \to 4\tilde{J}$, we arrive at our effective Hamiltonian Eq. (5.1) with coupling strength

$$\lim_{\substack{\kappa\to\infty \\ \gamma\neq 0,\ k<\infty,\ \Delta r<\infty}} \tilde{J} = \frac{\beta k^2 (h_0 - l_0)^2}{16\sqrt{\gamma\kappa}} + \mathcal{O}\left(\frac{\ln\kappa}{\kappa}\right). \tag{5.25}$$

Here we find the relation $\tilde{J} \propto 1/\sqrt{\kappa}$, as mentioned in the main text.

Appendix 3: Non-uniform Force Distribution

Under the condition of a small combined elastic modulus, corresponding to large values of the coupling strength $\tilde{J} \gg 1$, the assumption of an equally shared force load is no longer valid [35, 64, 65]. We therefore address how a **non-uniform force distribution** affects the EOS and MFPT to cluster dissolution/formation for finite system sizes. Based on Eq. (7) in [65] and Eq. (4) in [124] we introduce a non-uniform force load by making the substitution $h \to \mathcal{C}\sum_i h_i \delta_{\sigma_i,-1}/N_c(\{\sigma_i\})$ in Eq. (5.2), where $\mathcal{C} \equiv Nh/\sum_i h_i$ is a normalization constant such that initially, i.e. when all bonds are closed, the total force load is h. The load on bond i, denoted as h_i, is given by

$$h_i = \frac{1}{\sqrt{\xi - \bar{\epsilon}_i^2}}, \tag{5.26}$$

where $\xi \geq 1$, and $\bar{\epsilon}_i \equiv (\epsilon_i - r)/(d - r) \in [0, 1]$ is a normalized distance of bond i to the center of the lattice, with ϵ_i defined as the eccentricity of node i, which is the maximum number of edges between node i and any other node in the lattice. The radius $r \equiv \min \epsilon_i$ and diameter $d \equiv \max \epsilon_i$ of the lattice are defined as the minimum and maximum eccentricity, respectively. With the force distribution given by Eq. (5.26), which is depicted in Fig. 5.12a, closed bonds located at the outer edge of the lattice ($\epsilon_i = 1$) experience a larger external force than bonds located at the inner part of the lattice ($\epsilon_i = 0$). The parameter ξ is an indicator for the spread in force load among the individual bonds. For $\xi = 1$, which holds when $\lim \tilde{J} \to \infty$ [65], the force distribution at the edge of the cluster is singular and nonphysical. On the contrary, for $\lim \xi \to \infty$, which is valid for $\lim \tilde{J} \to 0$, we recover the uniform force distribution.

In Fig. 5.12b–d we depict the equation of state (b) and MFPT to cluster dissolution (c) and formation (d) for mixed Glauber-Kawasaki dynamics with a constant Glauber attempt probability $p = 0.5$ and for various values of ξ under a pulling or pushing force ($\tilde{h} = \pm 0.5$). The results were obtained by exact summation/algebraic techniques. Interestingly, for $\xi \geq 1.1$ the EOS and MFPT are almost identical to the uniform force load solutions corresponding to $\xi \to \infty$. Only for $\xi < 1.1$, corresponding to extremely floppy membranes, we observe deviations from the uniform force results. The origin of the deviations is the extreme force load on the outer bonds, which is $\sqrt{\xi/(\xi - 1)}$ times larger than the force load on the inner bond. For $\xi = 1.01$ this leads approximately to a factor of $\times 10$, and for $\xi = 1.001$ this leads approximately to a factor of $\times 32$. Hence, for most physically meaningful realizations of a non-uniform force distribution [i.e. distributions based on Eq. (5.26)] the results converge to the uniform force solutions. Only under the extreme conditions where the force load on the outer bonds becomes at least an order of magnitude larger compared to the inner bonds we find large deviations from the uniform force load. Note that the relative fraction of edge bonds in the thermodynamic limit vanishes. Therefore we expect a non-uniform force load, which mainly penalizes the edge bonds for $\xi \to 1$, to have an even weaker effect on the EOS and mean first passage times in large systems.

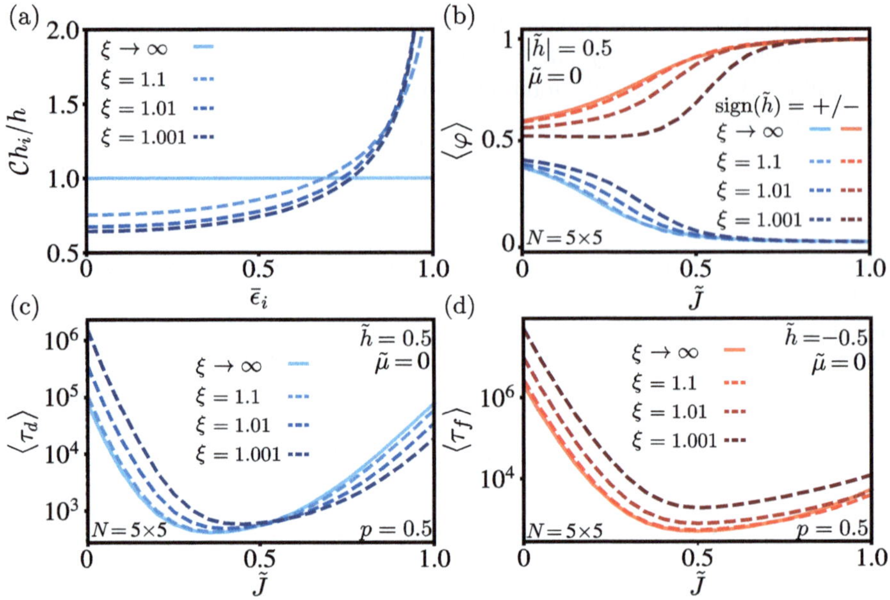

Fig. 5.12 Comparison between a uniform and non-uniform force load. a The non-uniform force load distribution given by Eq. (5.26) as a function of the normalized lattice distance $\bar{\epsilon}_i$. For $\xi < \infty$ the external force increases monotonically with the lattice distance. **b** The EOS for a pulling (blue) and pushing (red) force for various values of ξ for a system of $N = 5 \times 5$ adhesion bonds with zero intrinsic binding-affinity. **c–d** The MFPT to dissolution (**c**) and formation (**d**) for mixed Glauber-Kawasaki dynamics with constant Glauber attempt probability $p = 0.5$ under a pulling (blue) and pushing (red) force for various values of ξ and $N = 5 \times 5$ adhesion bonds with zero intrinsic binding-affinity

Appendix 4: Statical Critical Point

We determine the BG statical critical point from $\tilde{\mathrm{f}}'''_{\mathrm{BG}}(\varphi) = 0$, which results in the following equation using Eq. 5.10

$$\alpha_{\tilde{h}}\varphi^2 (1-\varphi)^2 + (1+\varphi)^4 (2\varphi - 1)\, \gamma_{\bar{z}}(\varphi) = 0, \tag{5.27}$$

with $\alpha_{\tilde{h}} \equiv 12\tilde{h}$, and we have furthermore defined the auxiliary function

$$\gamma_{\bar{z}}(\varphi) \equiv 1 - \frac{1}{\bar{z}}\left(3 - \frac{2}{\bar{z}}\right) + \frac{2\alpha_{\tilde{h}}}{\bar{z}}\left(1 - \frac{1}{\bar{z}}\right)\Lambda(\varphi) - \frac{\alpha_{\tilde{h}}^2}{3\bar{z}}\left(1 - \frac{2}{\bar{z}}\right)\Lambda^2(\varphi) - \frac{2\alpha_{\tilde{h}}^3}{27\bar{z}^2}\Lambda^3(\varphi), \tag{5.28}$$

with $\Lambda(\varphi) \equiv \varphi(1-\varphi)/(1+\varphi)^3$. For $\tilde{h} = 0$ the solution of Eq. (5.27) is given by $\varphi^{\mathrm{s}}_{\mathrm{crit,BG}} = 1/2$, and the corresponding statical critical point is given by $\tilde{J}^{\mathrm{s}}_{\mathrm{crit,BG}} = \ln(\bar{z}/(\bar{z}-2))/\bar{2}$ as shown in Sect. 2.8.3. For non-zero force we solve Eq. (5.27) by means of a **quadratic Newton series** as explained in more detail in the next Appendix 5. The main result for the statical critical fraction obtained by the quadratic Newton series reads

$$\varphi^{\mathrm{s}}_{\mathrm{crit}} \approx \frac{1}{2} - \frac{3}{2^4}\frac{\delta_{\bar{z}}(\tilde{h})}{\nu_{\bar{z}}(\tilde{h})} + \frac{3}{2^4}\left[\frac{\delta^2_{\bar{z}}(\tilde{h})}{\nu^2_{\bar{z}}(\tilde{h})} - \frac{2^6}{3^4}\frac{\bar{z}^2\tilde{h}}{\nu_{\bar{z}}(\tilde{h})}\right]^{\frac{1}{2}}, \tag{5.29}$$

where the auxiliary functions $\delta_{\bar{z}}(\tilde{h})$ and $\nu_{\bar{z}}(\tilde{h})$ are defined as

$$\delta_{\bar{z}}(\tilde{h}) \equiv \left(\bar{z} - 1 - \frac{2^6}{3^5}\tilde{h}^2\right)\left(\bar{z} - 2 + \frac{2^4}{3^2}\tilde{h}\right) + \frac{2^{13}}{3^9}\tilde{h}^3, \tag{5.30}$$

$$\nu_{\bar{z}}(\tilde{h}) \equiv \left(\bar{z} - 1 + \frac{2^5}{3^5}\tilde{h}^2\right)\left(\bar{z} - 2 + \frac{2^2}{3^2}\tilde{h}\right) - \frac{2}{3^2}\bar{z}^2\tilde{h} + \frac{2^7}{3^9}\tilde{h}^3, \tag{5.31}$$

respectively. The statical critical coupling $\tilde{J}^{\mathrm{s}}_{\mathrm{crit}}$ is obtained by inserting Eq. (5.29) into Eq. (5.12), and the result is depicted by the gradient line Fig. 5.6 in the main text, where the black symbols represent the fully converged Newton's series (5.33).

Appendix 5: Equation of State in the Thermodynamic Limit

Here we derive the EOS in the thermodynamic limit using the saddle-point technique, i.e.

$$\begin{aligned}\langle\varphi\rangle_{\mathrm{TD}} &\equiv \lim_{N\to\infty}\frac{1}{\mathcal{N}}\int_0^1 \varphi \mathrm{e}^{-N\tilde{\mathrm{f}}(\varphi)}\mathrm{d}\varphi \simeq \lim_{N\to\infty}\frac{1}{\mathcal{N}}\int_0^1 \sum_{i=1}^{M}\varphi_i^0 \mathrm{e}^{-N\tilde{\mathrm{f}}''(\varphi_i^0)(\varphi-\varphi_i^0)^2}\mathrm{d}\varphi \\ &= \sum_{i=1}^{M} c_i\varphi_i^0,\end{aligned}$$

where $\varphi_1^0, \ldots, \varphi_M^0$ denote the locations of the local minima of the BG free energy density $\tilde{\mathrm{f}}(\varphi) \equiv \tilde{\mathrm{f}}_{\mathrm{BG}}(\varphi)$ given by Eq. (5.10), $\mathcal{N} = \int_0^1 \exp\left(-N\tilde{\mathrm{f}}(\varphi)\right)d\varphi$, and the coefficients c_i read

$$c_j = \lim_{N\to\infty}\mathcal{N}^{-1}\int_0^1 \mathrm{e}^{-N\tilde{\mathrm{f}}''(\varphi_j^0)(\varphi-\varphi_j^0)^2}d\varphi \simeq \left(1 + \sum_{i=1|i\neq j}^{M}[\tilde{\mathrm{f}}''(\varphi_j^0)/\tilde{\mathrm{f}}''(\varphi_i^0)]^{1/2}\right)^{-1}.$$

The idea behind the above result is that in the large N limit we expect the integral over φ to be dominated by the immediate neighborhood of the local minima of $\tilde{\mathrm{f}}(\varphi)$. We may therefore approximate the exponent by its Taylor expansion around these extremal points. In general special care has to be taken when one of the global minima lies at the boundary of the integration interval [125], which turns out not to be the case here. For $\tilde{h} = 0$ the local minima of the BG free energy density are analyzed in Sect. 2.9. Here we proceed with the analysis for non-zero force $\tilde{h} \neq 0$ and $\bar{z} = 4$. Based on the analysis in Sect. 2.9.2 we obtain a transcendental equation for the auxiliary variable $\xi^4 \equiv \varphi/(1-\varphi)$ which reads

$$g(\xi) \equiv \xi^4 - \mathrm{e}^{2\tilde{J}}[\mathrm{e}^{\frac{\tilde{\mu}}{4}-\frac{1}{2}\left(\frac{1+\xi^4}{1+2\xi^4}\right)^2\tilde{h}}\xi^3 - \mathrm{e}^{-\frac{\tilde{\mu}}{4}+\frac{1}{2}\left(\frac{1+\xi^4}{1+2\xi^4}\right)^2\tilde{h}}\xi] - 1 = 0. \tag{5.32}$$

For $\tilde{h} = 0$ Eq. (5.32) reduces to a quartic similar to Eq. (2.29). For $\tilde{h} \neq 0$ we determine the roots of $g(\xi)$ by means of a convergent Newton series yielding the exact result [70]

$$\xi = \xi_0 - \sum_{k=1}^{\infty} \frac{\left[g^{(0)}(\xi_0)\right]^k}{\left[g^{(1)}(\xi_0)\right]^{2k-1}} \frac{\det \mathcal{A}_k(\xi_0)}{(k-1)!}, \tag{5.33}$$

where ξ_0 is an initial guess in a convex neighborhood of $g(\xi)$, $g^{(k)}(\xi_0) = d^k g(\xi) / d\xi^k|_{\xi=\xi_0}$, and $\mathcal{A}_k(\xi_0)$ are upper/lower Hessenberg matrices[3] of size $(k-1) \times (k-1)$ with elements

$$\mathcal{A}_k^{ij}(\xi_0) = \frac{g^{(i-j+2)}(\xi_0)\,\theta(i-j+1)}{(i-j+2)!}\left(k[i-j+1]\,\theta(j-2) + i\theta(1-j) + j - 1\right), \tag{5.34}$$

where $\theta(x)$ denotes the Heaviside step function. The determinant of Hessenberg matrices can be efficiently calculated using a recursion formula [126], for which a numerical implementation can be found in [127]. Yet it remains computationally expensive to evaluate the determinant in Eq. (5.33), and furthermore it does not provide any analytical insight. Therefore we now introduce the so-called quadratic approximation.

Appendix 5.1: Quadratic Newton Series

If we set $g^{(3)} = g^{(4)} = \cdots = 0$ in the almost triangular matrices, the resulting matrix $\tilde{\mathcal{A}}_k$ becomes triangular, implying that its determinant is simply given by the product of its diagonal elements. Making the substitution $\mathcal{A}_k \to \tilde{\mathcal{A}}_k$ in Eq. (5.33) yields the so-called quadratic approximation [72, 128]

$$\xi \approx \xi_0 - \frac{g^{(1)}(\xi_0) - \sqrt{g^{(1)}(\xi_0)^2 - 2g^{(0)}(\xi_0)\,g^{(2)}(\xi_0)}}{g^{(2)}(\xi_0)}. \tag{5.35}$$

that becomes exceedingly accurate when the root moves close to ξ_0. For the initial point ξ_0 we use the *Ansatz*

[3] These are almost triangular matrices with non-zero elements in the first sub- or superdiagonal.

$$\xi_0 = \exp\left(\frac{\tilde{\mu}}{4} - \frac{2}{9}\tilde{h} + \text{sign}\left[\frac{\tilde{\mu}}{4} - \frac{2}{9}\tilde{h}\right]\frac{6}{5}\tilde{J}\right), \tag{5.36}$$

which is derived by considering an adapted form of Eqs. (2.30a) and (2.30b) in combination with the implementation of the force term. The weight 2/9 is derived from the term $(1+\xi^4)^2/2(1+2\xi^4)^2$ in Eq. (5.32) evaluated at the point $\xi = 1^4$ and the weight 6/5 in front of $\tilde{J}$ was selected empirically. This choice assures that Eq. (5.32) satisfies the Lipschitz condition between ξ_0 and the root ξ for a broad range of parameter values, and thus assures the convergence of the Newton's series. Plugging Eq. (5.36) into Eq. (5.35), and using the relation $\varphi_1^0 = \xi^4/(1+\xi^4)$, we obtain the location of the global minimum—and thus $\langle\varphi\rangle_{\text{TD}}$—for non-zero force. Notably, the *Ansatz* given by Eq. (5.36) also provides a numerically correct solution for zero force and non-zero intrinsic binding-affinity. Below we write down explicitly the terms which are used to evaluate Eq. (5.35). Introducing the auxiliary functions

$$\alpha_{\tilde{h}}(\xi) \equiv 4\xi^4\left(\xi^4+1\right)\tilde{h}/\left(2\xi^4+1\right)^3, \quad \beta_{\tilde{h}}(\xi) \equiv 4\xi^4(10\xi^8+11\xi^4-3)\tilde{h}/\left(2\xi^4+1\right)^4,$$

the first and second derivative can be written as

$$g^{(1)}(\xi) = 4\xi^3 - e^{2\tilde{J}}\left(c_{\tilde{\mu},\tilde{h}}(\xi)\left[3+\alpha_{\tilde{h}}\right]\xi - c^{-1}_{\tilde{\mu},\tilde{h}}(\xi)\left[1-\alpha_{\tilde{h}}\right]\xi^{-1}\right)\xi, \tag{5.37}$$

$$g^{(2)}(\xi) = 12\xi^2 - e^{2\tilde{J}}\left(c_{\tilde{\mu},\tilde{h}}(\xi)\left[(3+\alpha_{\tilde{h}})^2 - 3 - \beta_{\tilde{h}}\right]\xi + c^{-1}_{\tilde{\mu},\tilde{h}}(\xi)[2\alpha_{\tilde{h}} - 3\alpha^2_{\tilde{h}} + \beta_{\tilde{h}}]\xi^{-1}\right), \tag{5.38}$$

where $c_{\tilde{\mu},\tilde{h}}(\xi) \equiv \exp\left(\tilde{\mu}/4 - (1+\xi^4)^2\tilde{h}/2(1+2\xi^4)^2\right)$. Equation (5.35) provides our main result for the EOS in the thermodynamic limit in the presence of an external force. In Fig. 5.5 we show the results for various values of the force and intrinsic binding-affinity.

Appendix 6: Kinetics of Cluster Formation and Dissolution

Appendix 6.1: Exact Algebraic Result for Small Clusters

It is well known that the transition matrix for an absorbing discrete-time Markov chain with a set of recurrent states has the canonical form [68]

$$\mathbf{P} = \begin{bmatrix} \mathbf{1} & \mathbf{0} \\ \mathbf{R} & \mathbf{T}_{\text{d,f}} \end{bmatrix}, \tag{5.39}$$

where $\mathbf{1}$ is the identity matrix, $\mathbf{T}_{\text{d,f}}$ is the submatrix of transient states in dissolution/formation, and $\mathbf{R}$ the submatrix of recurrent states. In the particular case of

[4] This corresponds to the point $\varphi = 1/2$.

cluster dissolution the $(2^N - 1) \times (2^N - 1)$ matrix $\mathbf{T}_\mathrm{d}$ entering Eq. (5.39) is obtained by removing the last column and row, and the $(2^N - 1) \times (2^N - 1)$ matrix $\mathbf{T}_\mathrm{f}$ entering Eq. (5.39) by removing the first column and row. If we introduce the column vector $\hat{\mathbf{e}}_k$ with components $(\hat{\mathbf{e}}_k)_i = \delta_{ki}$ and the column vector $\mathbf{e}$ whose elements are all equal to 1, the mean first passage times for cluster formation and dissolution read exactly

$$\langle \tau_d \rangle = \hat{\mathbf{e}}_1^T (\mathbf{1} - \mathbf{T}_\mathrm{d})^{-1}\mathbf{e}, \quad \langle \tau_f \rangle = \hat{\mathbf{e}}_{2^N-1}^T (\mathbf{1} - \mathbf{T}_\mathrm{f})^{-1}\mathbf{e}. \tag{5.40}$$

To apply Eq. (5.40) one must invert a $(2^N-1)\times(2^N-1)$ sparse matrix and afterwards sum over 2^N-1 terms, which is feasible for $N \lesssim 25$. For $N = 20$ bonds the exact results are shown in Fig. 5.7b, c (blue line). Larger clusters are treated within the *LEQ approximation*.

Appendix 6.2: Proof of DB for the LEQ Transition Rates

Before stating the explicit result for the MFPT to dissolution/formation within the LEQ approximation, we prove that the LEQ transition rates $\bar{w}_{k\to k\pm 1}$ given by Eq. (5.17) obey DB w.r.t. $\hat{\mathcal{Z}}_k$ defined in Eq. (5.6). The effective transition rates are obtained by mapping the full mixed Glauber-Kawasaki dynamics onto an effective birth-death process over the number of closed bonds (see Fig. 5.7a), where we assume that the dynamics reaches a LEQ at any number of closed bonds before any transition. As a result, the birth-death process is a Markov chain on the free energy landscape for the fraction of closed bonds φ. Recall that the original Glauber rates in Eq. (5.3) obey DB w.r.t. the Hamiltonian $\mathcal{H}(\{\sigma_j\})$, hence

$$\begin{aligned} &\mathrm{e}^{-\beta\mathcal{H}(\{\sigma_j\})} w_i(\{\sigma_j\})\delta_{N_c(\{\sigma_j\}),k}\delta_{N_c(\{\sigma_j\}_i'),k\pm 1} \\ &= \mathrm{e}^{-\beta\mathcal{H}(\{\sigma_j\}_i')} w_i(\{\sigma_j\}_i')\delta_{N_c(\{\sigma_j\}_i'),k\pm 1}\delta_{N_c(\{\sigma_j\}),k} \end{aligned} \tag{5.41}$$

where we have explicitly incorporated the constraints arising from single-bond-flip dynamics by means of the Kronecker delta's. Upon summing the LHS of Eq. (5.41) over all initial configurations with $N_c(\{\sigma_j\}) = k$ and over all rates that jump to a configuration with $N_c(\{\sigma_j\}_i') = k \pm 1$, we reach all possible final configurations with $N_c(\{\sigma_j\}_i') = k \pm 1$, with a backward rate given by the sum of all rates that jump to a configuration with $N_c(\{\sigma_j\}) = k$. Hence we obtain the equality

$$\begin{aligned} &\sum_{\{\sigma_j\}}\sum_{i=1}^{N} \mathrm{e}^{-\beta\mathcal{H}(\{\sigma_j\})} w_i(\{\sigma_j\})\delta_{N_c(\{\sigma_j\}),k}\delta_{N_c(\{\sigma_j\}_i'),k\pm 1} = \\ &\sum_{\{\sigma_j\}_i'}\sum_{i=1}^{N} \mathrm{e}^{-\beta\mathcal{H}(\{\sigma_j\}_i')} w_i(\{\sigma_j\}_i')\delta_{N_c(\{\sigma_j\}_i'),k\pm 1}\delta_{N_c(\{\sigma_j\}),k}. \end{aligned} \tag{5.42}$$

Comparing Eq. (5.42) with Eq. (5.17), we recognize the LHS and RHS as $(\hat{\mathcal{Z}}_k/p_k)$ $\bar{w}_{k\to k\pm 1}$ and $(\hat{\mathcal{Z}}_{k\pm 1}/p_{k\pm 1})\bar{w}_{k\pm 1\to k}$, respectively, which proves Eq. (5.42).

Appendix 6.3: FPT Statistics Within the LEQ Approximation

The LEQ approximation maps the complete mixed Glauber-Kawasaki dynamics onto an effective birth-death process with a right-acting tri-diagonal transition matrix $\mathbf{P}^{\text{le}}$ of size $(N+1)\times(N+1)$ with elements

$$P_{ij}^{\text{le}} = \Lambda_i\delta_{ij} + \bar{w}_{i-1\to i}\delta_{i+1j}\theta\,(N-i) + \bar{w}_{i-1\to i-2}\delta_{i-1j}\theta\,(i-2)\,, \tag{5.43}$$

and $\Lambda_i = 1-\sum_{j\neq i}^{N+1} P_{ij}^{\text{le}}$. To obtain the MFPT we use the same algebraic technique as for small clusters. Upon removing the first/last row and column of $\mathbf{P}^{\text{le}}$ we obtain the submatrix $\mathbf{T}_{\text{d,f}}^{\text{le}}$ for cluster dissolution and formation, respectively. We can invert the tri-diagonal submatrix exactly, which leads to the following LU/UL decomposition

$$(\mathbf{1}-\mathbf{T}_{\text{d,f}}^{\text{le}})^{-1} = \mathcal{A}^{\text{d,f}}\mathcal{B}^{\text{d,f}}, \tag{5.44}$$

where $\mathcal{A}^{\text{d}}$ and $\mathcal{B}^{\text{d}}$ are the lower and upper triangular matrix with elements

$$\mathcal{A}_{ij}^{\text{d}} = (p_{j-1}/\hat{\mathcal{Z}}_{j-1})\bar{w}_{j-1\to j}\theta\,(i-j)\,,\quad \mathcal{B}_{ij}^{\text{d}} = (\hat{\mathcal{Z}}_j/p_j)\theta\,(j-i)\,, \tag{5.45}$$

and $\mathcal{A}^{\text{f}}$ and $\mathcal{B}^{\text{f}}$ are the upper and lower triangular matrix with elements

$$\mathcal{A}_{ij}^{\text{f}} = (p_j/\hat{\mathcal{Z}}_j)\bar{w}_{j\to j-1}\theta\,(j-i)\,,\quad \mathcal{B}_{ij}^{\text{f}} = (\hat{\mathcal{Z}}_{j-1}/p_{j-1})\theta\,(i-j)\,. \tag{5.46}$$

A proof of Eq. (5.44) is given in the SM of [1]. Let us denote with $\langle\tau_{d,f}^{\text{le}}\rangle_m$ the MFPT to cluster dissolution and formation, starting from the state with m closed bonds. Using Eq. (5.44) we obtain an exact expression for the first moments

$$\langle\tau_d^{\text{le}}\rangle_{0<m\leq N} = \hat{\mathbf{e}}_m^T\mathcal{A}^{\text{d}}\mathcal{B}^{\text{d}}\mathbf{e} = \sum_{k=0}^{m-1}\frac{1}{\bar{w}_{k\to k+1}}\sum_{l=k+1}^{N}\frac{\hat{\mathcal{Z}}_l}{\hat{\mathcal{Z}}_k}\frac{p_k}{p_l}, \tag{5.47}$$

$$\langle\tau_f^{\text{le}}\rangle_{0\leq m<N} = \hat{\mathbf{e}}_{m+1}^T\mathcal{A}^{\text{f}}\mathcal{B}^{\text{f}}\mathbf{e} = \sum_{k=m+1}^{N}\frac{1}{\bar{w}_{k\to k-1}}\sum_{l=0}^{k-1}\frac{\hat{\mathcal{Z}}_l}{\hat{\mathcal{Z}}_k}\frac{p_k}{p_l}, \tag{5.48}$$

where $\hat{\mathbf{e}}_m$ is the column vector with dimension N with components $(\hat{\mathbf{e}}_m)_i = \delta_{mi}$, and $\mathbf{e}$ is the column vector with all components equal to 1. Notice that Eq. (5.44), and therefore Eqs. (5.47) and (5.48), are applicable to any right-acting tri-diagonal transition matrix with rates obeying DB. Although we only present the MFPT here, we can easily obtain any higher order moments of the first passage time to cluster dissolution and formation using Eq. (5.44) [68]. Notably, Eqs. (5.47) and (5.48) appear to have a similar structure as the largest eigenvalue of the transition matrix in classical nucleation theory [110, 129].

Appendix 6.4: A Bound on the Effective Transition Rates

Here we present a bound on the LEQ rates given by Eq. (5.17) which proves that the transition rates are strictly sub-exponential in N. First, we consider a bound for the exit rates $w^{\pm}_{\mathrm{exit}}(\{\sigma_i\})$ defined in Eq. (5.18), which contain a sum over the original Glauber rates that are defined in Eq. (5.3). Since $1 - \tanh(x) \geq 0,\ \forall x \in \mathbb{R}$ the Glauber rates are non-negative, and therefore the exit rates obey the bound

$$w^{\max}_{k\to k\pm 1} \leq w^{\pm}_{\mathrm{exit}}(\{\sigma_i\}) \leq c^{\pm}_k w^{\max}_{k\to k\pm 1}, \tag{5.49}$$

with $c^{+}_k = N - k$ and $c^{-}_k = k$ denoting the number of terms inside the sum of Eq. (5.18), and $w^{\max}_{k\to k\pm 1}$ denotes the largest transition rate to go from a state with k to $k \pm 1$ closed bonds. The largest transition rate can be written as

$$w^{\max}_{k\to k\pm 1} = \frac{1}{2N}\left[1 - \tanh\left(\Delta\mathcal{H}^{\min}_{k\to k\pm 1}/2\right)\right],$$

where

$$\Delta\mathcal{H}^{\min}_{k\to k\pm 1} \equiv \inf_{\substack{N_c(\{\sigma_j\})=k,\\ N_c(\{\sigma_j\}'_i)=k\pm 1}} \{\mathcal{H}(\{\sigma_j\}'_i) - \mathcal{H}(\{\sigma_j\})\} \tag{5.50}$$

denotes the smallest possible energy change between two configurations $\{\sigma_j\}$ and $\{\sigma_j\}'_i$ with $N_c(\{\sigma_j\}) = k$ and $N_c(\{\sigma_j\}'_i) = k \pm 1$, respectively. To obtain a closed-form expression for $w^{\max}_{k\to k\pm 1}$ we first note that the contribution to $\Delta\mathcal{H}^{\min}_{k\to k\pm 1}$ from the external force and intrinsic binding-affinity are fixed and given by the second and third term in Eq. (5.3). Therefore we are left to consider the smallest energy change due to the coupling strength, which we denote as $\Delta_{\tilde{J}}\mathcal{H}^{\min}_{k\to k\pm 1}$. For a square lattice with free boundary conditions the minimal energy "forward transitions" with energy difference $\Delta_{\tilde{J}}\mathcal{H}^{\min}_{k\to k+1}$ for various values of k are depicted in Fig. 5.13. Similarly, the minimal energy "downward transitions" with energy difference $\Delta_{\tilde{J}}\mathcal{H}^{\min}_{k\to k-1}$ are obtained by interchanging the open (red) and closed (green) adhesion pairs in Fig. 5.13. Combining these two results yields $\Delta_{\tilde{J}}\mathcal{H}^{\min}_{k\to k\pm 1} = 2m^{\pm}_k \tilde{J}$, with

$$m^{\pm}_k \equiv 2(c^{\mp}_k - 1)\theta(2 - c^{\mp}_k) - \min(c^{\mp}_k, 4)\theta(c^{\mp}_k - 3),$$

and delivers the expression for $w^{\max}_{k\to k\pm 1}$. Finally, since $w^{\max}_{k\to k\pm 1}$ is independent of the specific configuration $\{\sigma_i\}$ at fixed k, it drops out of the sum over $\{\sigma_i\}$ in Eq. (5.17)

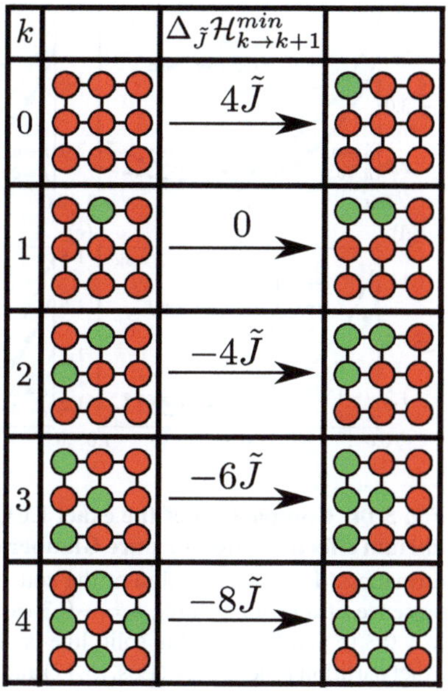

Fig. 5.13 Minimum energy forward transitions between two configurations $\{\sigma_j\}$ and $\{\sigma_j\}'_i$ with $N_c(\{\sigma_j\}) = k$ and $N_c(\{\sigma_j\}'_i) = k+1$ respectively. Although we depict here the minimal energy differences for a lattice of size $N = 3\times3$, the result holds for any two-dimensional lattice of size $N \geq 3\times3$ as long as the transitions for $k \leq 2$ are taken at the corner, and the transition for $k = 3$ is taken at the edge

for the effective transition rates, and therefore the bound in Eq. (5.49) can directly be applied to the effective transition rate upon multiplying both sides with the Glauber attempt probability

$$p_k w^{\max}_{k\to k\pm1} \leq \bar{w}_{k\to k\pm1} \leq p_k c^{\pm}_k w^{\max}_{k\to k\pm1}, \tag{5.51}$$

which yields the bound on the effective transition rate. The lower and upper bound for the effective transition rate are used to determine an upper and lower bound for the MFPT to cluster dissolution and formation, respectively. The specific result for a lattice of size $N = 6 \times 7$ with $\bar{z} = 4$ for pure Glauber dynamics (i.e. $p_k = 1,\ \forall k$) is shown in Fig. 5.14. For small values of the coupling strength $\tilde{J}$ we find that the upper bound in Eq. (5.51), corresponding to the lower bound in Fig. 5.14 (since $\langle \mathrm{t_{d,f}}\rangle \propto 1/\bar{w}_{k\to k\pm1}$), is saturated by the exact effective transition rate. Conversely, for large values of the coupling strength $\tilde{J}$ it seems that the lower bound in Eq. (5.51) is saturated.

Appendix 6.5: Approximate Effective Transition Rates

For systems larger than $N \approx 50$ bonds the combinatorics involved in the computation of $\hat{\mathcal{Z}}_k$ defined in Eq. (5.6) and $\bar{w}_{k\to k+1}$ in Eq. (5.17) becomes prohibitive, and thus forces us to make further approximations. To get Eq. (5.19) fully explicit we make an **instanton approximation** for $\bar{w}_{k\to k\pm1}$ using the BG approximation with the bound given by Eq. (5.51), and reads

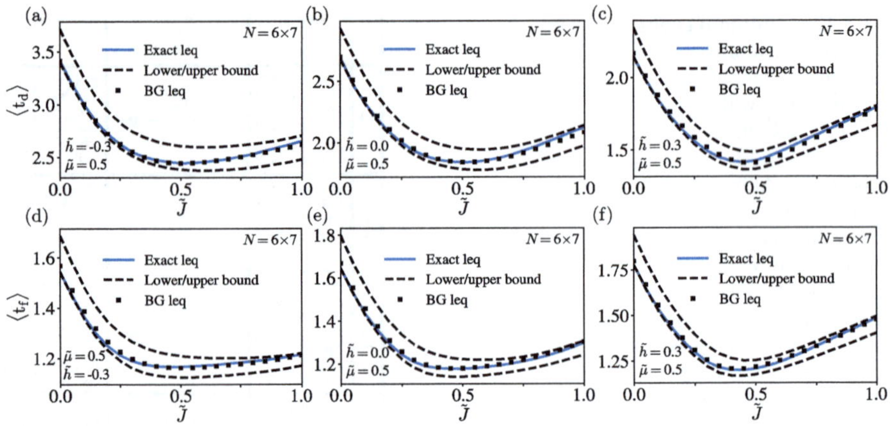

Fig. 5.14 Comparison of the exact LEQ effective rates and the approximate BG local equilibrium effective rates for pure Glauber dynamics. $\langle t_{d,f}\rangle$ for cluster dissolution (**a**–**c**) and cluster formation (**d**–**f**) as a function of the coupling $\tilde{J}$ for fixed intrinsic binding-affinity $\tilde{\mu} = 0.5$ in the presence of a pushing force $\tilde{h} = -0.3$ (**a** and **d**), zero force $\tilde{h} = 0$ (**b** and **e**), and a pulling force $\tilde{h} = 0.3$ (**c** and **f**); The blue solid line is obtained with the exact LEQ effective transition rate and exact partition function $\hat{\mathcal{Z}}_k$ (Eqs. (5.17) and (5.6), respectively) with $p_k = 1,\ \forall k$. The black symbols are obtained with the BG approximation to the effective rate and partition function $\hat{\mathcal{Z}}_k^{\mathrm{BG}}$ (Eqs. (5.17) and (5.6) in combination with Eq. (5.8), respectively). The black dotted line indicates the upper and lower bound to the mean dissolution/formation time, which is obtained with the upper and lower bound to the effective transition rate in combination with the exact partition function (Eqs. (5.51) and (5.6) respectively)

$$\bar{w}_{k\to k\pm1} \approx \max(1, \alpha_k c_k^{\pm}) p_k w_{k\to k\pm1}^{\max}, \tag{5.52}$$

with $\alpha_k = \delta_{k0} + \delta_{kN} + 2\zeta_{\mathrm{BG}}^{\dagger}(\varphi = k/N) \in [0, 1]$, $c_k^+ = N - k$, $c_k^- = k$, and $\zeta_{\mathrm{BG}}^{\dagger}(\varphi)$ given by Eq. (2.12). The prefactor $\alpha_k c_k^{\pm}$ is a measure for the number of favorable adhesion bonds that are most likely to flip in a configuration with k closed bonds. For $k = 0 \vee N$ all bonds have an equal surrounding in the thermodynamic limit (or for a periodic lattice), and therefore all $c_k^{\pm}$ open/closed bonds are equally likely to attempt a flip. For $0 < k < N$ it becomes energetically more favorable to flip a bond which is part of an open-closed adhesion pair (see Fig. 5.13). To determine the number of bonds that constitute an open-closed pair, we note that $\bar{z}N\zeta_{\mathrm{BG}}^{\dagger}(\varphi = k/N)$ is a measure for the number of open-closed pairs in a lattice of size N with k closed bonds. Upon dividing by the total number of pairs in the system, given by $\bar{z}N/2$, we obtain the probability $2\zeta_{\mathrm{BG}}^{\dagger}(\varphi = k/N) \in [0, 1]$ to select an open-closed pair in the lattice. Multiplying the result by the total number of open/closed adhesion bonds, i.e. $2c_k^{\pm}\zeta_{\mathrm{BG}}^{\dagger}(\varphi = k/N)$, we obtain an approximate expression for the number of open/closed bonds which constitute an open-closed adhesion pair.

To prove that the approximate effective rate given by Eq. (5.52) obeys the bound given by Eq. (5.51) we apply a chain of inequalities. First we note that $0 \leq 2\zeta_{\mathrm{BG}}^{\dagger}(\varphi) \leq 1/2$, where the upper bound follows from considering $\tilde{J} = 0$ and $\varphi = 1/2$

in Eq. (2.12), and the lower bound is given for $\varphi = 0 \vee 1$ or the limit $\tilde{J} \to \infty$. From this it follows that $0 \leq \alpha_k \leq 1$, and therefore $1 \leq \max(1, \alpha_k c_k^{\pm}) \leq c_k^{\pm}$. Finally, since we use $p_k w_{k\to k\pm 1}^{\max}$ in Eq. (5.52), it cancels on both sides of the inequality in Eq. (5.51), which leaves to prove the inequality we have proven above and thereby completes the proof.

Figure 5.14 shows the MFPT to cluster dissolution and formation obtained with the approximate effective rates (5.52) in combination with the BG approximation for $\hat{\mathcal{Z}}_k$ for a lattice of size $N = 6 \times 7$ for pure Glauber dynamics (i.e. $p_k = 1,\ \forall k$). The results obtained with the approximate rates (black symbols) agree to a high degree with the results obtained by the exact effective rates (blue solid line).

Appendix 6.6: MFPT in the Thermodynamic Limit

Here we prove the result for the MFPT to dissolution/formation in the thermodynamic limit given by Eqs. (5.20) and (5.21) based on the LEQ approximation. Intuitively $\langle \tau_{d,f} \rangle$ must scale as $\langle \tau_{d,f} \rangle \sim \mathrm{e}^{N\Delta\tilde{\mathrm{f}}^{\dagger}}$, where $\Delta\tilde{\mathrm{f}}^{\dagger} \equiv \tilde{\mathrm{f}}_{\max} - \tilde{\mathrm{f}}_{\min}$ denotes the difference in the free energy density between the minimum, $\tilde{\mathrm{f}}_{\min} = \inf_{\varphi} \tilde{\mathrm{f}}(\varphi)$, and the maximum, $\tilde{\mathrm{f}}_{\max} \equiv \sup_{\varphi} \tilde{\mathrm{f}}(\varphi)$, that for large clusters becomes independent of N. Indeed, according to Eq. (5.19) we have $(p_k/p_l)(\hat{\mathcal{Z}}_l/\hat{\mathcal{Z}}_k) = (p_k/p_l)\mathrm{e}^{N[\tilde{f}_N(k/N)-\tilde{f}_N(l/N)]}$ and recall that $\bar{w}_{k\to k+1}$ is strictly sub-exponential in N. Furthermore we make the assumption that the Glauber attempt probabilities p_k are *strictly sub-exponential* in N. Since both series in Eq. (5.19) are absolutely convergent, we can apply a version of the squeeze theorem to Eq. (5.19).

To simplify notation we write the summands in Eq. (5.19) as $0 < (p_k/p_l)a_{k,l}/\bar{w}_{k\to k+1} < \infty$, where $a_{k,l} \equiv \mathrm{e}^{N[\tilde{f}_N(k/N)-\tilde{f}_N(l/N)]}$. If $k^{\dagger}$ denotes the index of the largest k-dependent term

$$k^{\dagger} \equiv \sup_{0\leq k<N} \frac{p_k \exp[N\tilde{f}_N(k/N)]}{\bar{w}_{k\to k+1}}, \tag{5.53}$$

and $l_{d,f}^{\dagger}$ the index of the largest l-dependent term

$$l_d^{\dagger} \equiv \sup_{k^{\dagger}<l\leq N} \frac{\exp[-N\tilde{f}_N(l/N)]}{p_l}, \quad l_f^{\dagger} \equiv \sup_{0<l<k^{\dagger}} \frac{\exp[-N\tilde{f}_N(l/N)]}{p_l}, \tag{5.54}$$

then the following chain of inequalities holds for any N

$$\frac{p_{k^{\dagger}} a_{k^{\dagger},l_{d,f}^{\dagger}}}{p_{l_{d,f}^{\dagger}} \bar{w}_{k^{\dagger}\to k^{\dagger}+1}} \leq \sum_{k=0}^{N-1}\sum_{l=m}^{M-1} \frac{p_k a_{k,l}}{p_l \bar{w}_{k\to k+1}} \leq \frac{c_{M,m}\, p_{k^{\dagger}} a_{k^{\dagger},l_{d,f}^{\dagger}}}{p_{l_{d,f}^{\dagger}} \bar{w}_{k^{\dagger}\to k^{\dagger}+1}}, \tag{5.55}$$

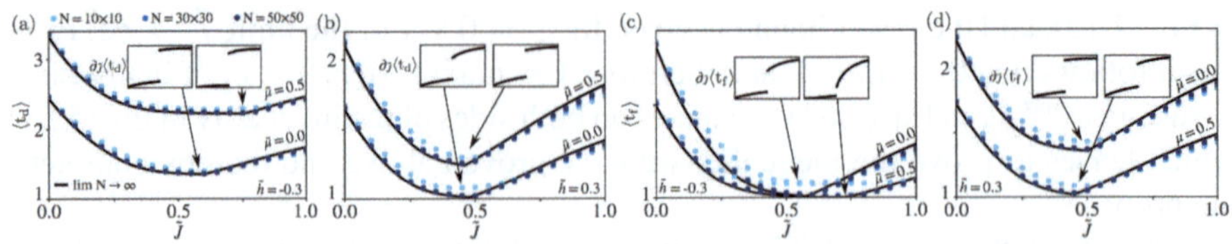

Fig. 5.15 Master scaling of mean dissolution and formation times per bond for finite clusters and in the thermodynamic limit. $\langle t_{d,f}\rangle$ for cluster dissolution (**a, b**) and formation (**c, d**) as a function of the coupling $\tilde{J}$ for a pair of intrinsic affinities $\tilde{\mu}=0$ and $\tilde{\mu}=0.5$ and various cluster sizes (symbols) as well as the thermodynamic limit (lines) in the presence of an external pushing (**a, c**) and pulling (**b, d**) force; Symbols are evaluated with LEQ approximation Eq. (5.19) using $\hat{\mathcal{Z}}_k^{\mathrm{BG}}$ [Eqs. (5.6) and (5.8)] and $\bar{w}_{k\to k+1}$ from Eq. (5.52) with $p_k=1$, $\forall k$ (i.e. pure Glauber dynamics) The discrepancy between the lines and symbols is due to finite-size effects.

where $c_{M,m}\equiv N(M-m)$, $M=N+1$ and $m=k+1$ for dissolution, and $M=k+1$ and $m=0$ for cluster formation. Since $x^{1/N}$ is monotonic in $x>0$, such that $x_1<x_2$ implies $x_1^{1/N}<x_2^{1/N}$, the inequality (5.55) is preserved when exponentiated to $1/N$. The thermodynamic limit of Eq. (5.55) is a scaling limit, i.e. $\lim_s\equiv\lim_{N\to\infty}|_{k/N=\varphi_k}^{l/N=\varphi_l}$, and thus (5.55) becomes

$$\lim_s\left[\frac{p_{k^\dagger}a_{k^\dagger,l^\dagger_{d,f}}}{p_{l^\dagger_{d,f}}\bar{w}_{k^\dagger\to k^\dagger+1}}\right]^{\frac{1}{N}}\leq\langle t_{d,f}\rangle\leq\lim_s\left[\frac{c_{M,m}p_{k^\dagger}a_{k^\dagger,l^\dagger_{d,f}}}{p_{l^\dagger_{d,f}}\bar{w}_{k^\dagger\to k^\dagger+1}}\right]^{\frac{1}{N}}. \tag{5.56}$$

Moreover, since $\lim_s[\bar{w}_{k^\dagger\to k^\dagger+1}]^{-1/N}=1$, $\lim_s[c_{M,m}]^{1/N}=1$, and $\lim_s[p_{k^\dagger}/p_{l^\dagger_{d,f}}]^{1/N}=1$, all limits in Eq. (5.56) exist and may be taken separately, implying the convergence of the upper bound to the lower bound. Thereby $\langle t_{d,f}\rangle$ becomes *squeezed* in-between the upper and lower bound, rendering the inequality an equality. Since $\lim_s\tilde{f}_N(k^\dagger/N)=\tilde{f}(\varphi_{\max})$ and $\lim_s\tilde{f}_N(l^\dagger_{d,f}/N)=\tilde{f}(\varphi^{d,f}_{\min})$ we finally obtain Eqs. (5.20) and (5.21), completing the proof.

Appendix 6.7: BG MFPT in the Thermodynamic Limit

In the previous section we have proven that in the thermodynamic limit the MFPT to cluster dissolution/formation scales as $\langle\tau_{d,f}\rangle\simeq\langle t_{d,f}\rangle^N=e^{N\Delta\tilde{f}^\dagger}$, where $\Delta\tilde{f}^\dagger$ denotes the largest left/right barrier in the BG free energy density Eq. (5.10). In this section we determine $\Delta\tilde{f}^\dagger$ and thereby obtain a closed-form expression for the MFPT in the thermodynamic limit.

Case 1: $\tilde{J}\geq 0, \tilde{h}=\tilde{\mu}=0$

We first consider the MFPT to cluster dissolution/formation in the absence of an external force and intrinsic binding-affinity. Due to the $\mathbb{Z}_2$ symmetry of the coupling strength, we note that $\langle t_d\rangle=\langle t_f\rangle$. Our first task is to find the locations of the global maximum and minimum in the free energy landscape, denoted by $\varphi^{d,f}_{\max}$ and $\varphi^{d,f}_{\min}$, respectively.

The position of the global minimum for zero force and intrinsic binding-affinity is given by Eq. (2.37), while the position of the global maximum is located at $\varphi^{d,f}_{\max} = 0 \wedge 1$ for low values of the coupling strength $\tilde{J}$, and at $\varphi^{d,f}_{\max} = 1/2$ for large values. The coupling strength at which the global maximum changes position corresponds to the root of the equation

$$\tilde{\mathrm{f}}\,(0) - \tilde{\mathrm{f}}\,(1/2) = (\bar{z}/2)\ln\left(\mathrm{e}^{2\tilde{J}} + 1\right) - \bar{z}\tilde{J} + (1 - \bar{z}/2)\ln(2), \tag{5.57}$$

which is given by

$$\tilde{J}^{\mathrm{d}}_{\mathrm{crit,BG}} = -(1/2)\ln\left(2^{1-2/\bar{z}} - 1\right), \tag{5.58}$$

and sets the BG dynamical critical coupling value for the zero-field Ising model. Surprisingly, for the two-dimensional square lattice with $\bar{z} = 4$ we exactly recover the critical point obtained by Onsager [73]. To check whether this is a mere coincidence we note that for the honeycomb lattice with $\bar{z} = 3$ the exact statical critical point is given by $\tilde{J}^{\mathrm{s}}_{\mathrm{crit}} = \frac{1}{2}\ln\left(2 + \sqrt{3}\right) = 0.65\ldots$ [130], whereas Eq. (5.58) gives $\tilde{J}^{\mathrm{d}}_{\mathrm{crit}} = \frac{-1}{2}\ln\left(2^{1/3} - 1\right) = 0.67\ldots$, and so we find that the surprising equality for $\bar{z} = 4$ is coincidental.

Combining our results for the locations of the global maximum and minimum we obtain the following result for the MFPT per adhesion bond in the thermodynamic limit for the zero field Ising model with $\bar{z} = 4$

$$\ln\langle \mathrm{t_{d,f}}\rangle = \begin{cases} \tilde{\mathrm{f}}(0) - \tilde{\mathrm{f}}(\frac{1}{2}), & 0 \le \tilde{J} \le \tilde{J}^{\mathrm{s}}_{\mathrm{crit,BG}} \\ \tilde{\mathrm{f}}(0) - \tilde{\mathrm{f}}(\frac{1}{2}[1 \pm \mathcal{C}]), & \tilde{J}^{\mathrm{s}}_{\mathrm{crit,BG}} \le \tilde{J} \le \tilde{J}^{\mathrm{d}}_{\mathrm{crit,BG}} \\ \tilde{\mathrm{f}}(\frac{1}{2}) - \tilde{\mathrm{f}}(\frac{1}{2}[1 \pm \mathcal{C}]), & \tilde{J} \ge \tilde{J}^{\mathrm{d}}_{\mathrm{crit,BG}}, \end{cases} \tag{5.59}$$

where $\mathcal{C} \equiv \mathrm{e}^{2\tilde{J}}(\mathrm{e}^{4\tilde{J}} - 4)^{1/2}/(\mathrm{e}^{4\tilde{J}} - 2)$ comes from Eq. (2.37), $\tilde{J}^{\mathrm{s}}_{\mathrm{crit}} = \ln(2)/2$ denotes the statical critical point for zero force, and $\tilde{J}^{\mathrm{d}}_{\mathrm{crit}} = \ln(1 + \sqrt{2})/2$ the dynamical critical point in the force free case. This ultimately leads to Eq. (5.22) in the main text. In the strong coupling limit we find from Eq. (5.59) $\lim_{\tilde{J}\to\infty}\langle \mathrm{t_{d,f}}\rangle = 2$, which is identical to the result obtained for zero coupling. The physical intuition behind this comes from considering the average number of steps required to change the state of a single independent adhesion bond. For zero force and intrinsic binding-affinity the probability to associate/dissociate a bond is a $1/2$, and therefore the average dissolution/formation time is given by

$$1\left(\frac{1}{2}\right) + 2\left(\frac{1}{2}\right)^2 + 3\left(\frac{1}{2}\right)^3 + \ldots = \sum_{n=1}^{\infty} n\left(\frac{1}{2}\right)^n = 2.$$

For infinite coupling strength the interaction between the bonds is so strong that effectively the system behaves as one **super bond**, and therefore the average dissolution/formation time is equal to that of a single independent adhesion bond.

Case 2: $\tilde{J} \geq 0, \tilde{\mu} \neq 0, \tilde{h} = 0$

Here we use the results obtained in Sect. 2.9.2 which leads to the following outcomes

$$\ln\langle \mathrm{t_d} \rangle = \begin{cases} \tilde{\mathrm{f}}(0) - \tilde{\mathrm{f}}(\varphi_1), & 0 \leq \tilde{J} \leq \ln\min\{\sqrt{2}, s_0\} \\ \tilde{\mathrm{f}}(0) - \tilde{\mathrm{f}}(\varphi_3), & \ln\min\{\sqrt{2}, s_0\} \leq \tilde{J} \leq \tilde{J}^{\mathrm{d},-}_{\mathrm{crit}}, \\ \tilde{\mathrm{f}}(\varphi_4) - \tilde{\mathrm{f}}(\varphi_3), & \tilde{J} \geq \tilde{J}^{\mathrm{d},-}_{\mathrm{crit}} \end{cases} \tag{5.60}$$

$$\ln\langle \mathrm{t_f} \rangle = \begin{cases} \tilde{\mathrm{f}}(1) - \tilde{\mathrm{f}}(\varphi_1), & 0 \leq \tilde{J} \leq \ln\min\{\sqrt{2}, s_0\} \\ \tilde{\mathrm{f}}(1) - \tilde{\mathrm{f}}(\varphi_3), & \ln\min\{\sqrt{2}, s_0\} \leq \tilde{J} \leq \tilde{J}^{\mathrm{d},+}_{\mathrm{crit}}, \\ \tilde{\mathrm{f}}(\varphi_4) - \tilde{\mathrm{f}}(\varphi_1), & \tilde{J} \geq \tilde{J}^{\mathrm{d},+}_{\mathrm{crit}} \end{cases} \tag{5.61}$$

with $\varphi_i = \xi_i^4/(1+\xi_i^4)$ given by Eqs. (2.30a) and (2.30b), $s_0 \equiv \sqrt{2\sqrt{2}}\mathrm{e}^{-\tilde{\mu}/4}$, and $\tilde{J}^{\mathrm{d},\pm}_{\mathrm{crit}}$ are the dynamical critical points for cluster dissolution and formation, respectively given by

$$[\tilde{\mathrm{f}}(0) - \tilde{\mathrm{f}}(\varphi_4)]|_{\tilde{J}^{\mathrm{d},-}_{\mathrm{crit}}} \overset{!}{=} 0, \quad [\tilde{\mathrm{f}}(1) - \tilde{\mathrm{f}}(\varphi_3) - \tilde{\mathrm{f}}(\varphi_4) + \tilde{\mathrm{f}}(\varphi_1)]|_{\tilde{J}^{\mathrm{d},+}_{\mathrm{crit}}} \overset{!}{=} 0. \tag{5.62}$$

Case 3: $\tilde{J} \geq 0, \tilde{\mu} \neq 0, \tilde{h} \neq 0$

Using the quadratic Newton series (see Appendix 5.1), Eqs. (5.60) and (5.61) are directly applicable to the non-zero force scenario upon applying the transformation $\varphi_i \to \varphi_i^*$, where $\varphi_i^* = \xi_i^{*4}/(1+\xi_i^{*4})$ and

$$\xi_i^* = \xi_i - \frac{g^{(1)}(\xi_i) \pm \sqrt{g^{(1)}(\xi_i)^2 - 2g^{(0)}(\xi_i)\, g^{(2)}(\xi_i)}}{g^{(2)}(\xi_i)}, \tag{5.63}$$

with a minus sign for the global minimum $\xi^*_{1,3}$, and a plus sign for the global maximum ξ^*_4. The function $g^{(0)}(\xi)$ and its first and second derivative $g^{(1,2)}(\xi)$ are given in Eqs. (5.32),(5.37), and (5.38), respectively.

Our analytical results for the MFPT to cluster dissolution and formation per adhesion bond are depicted in Figs. 5.8 and 5.15 for zero and nonzero external force respectively; note the remarkable agreement between the black solid line depicting the thermodynamic limit and the results for finite system sizes on the order of $N \geq 10 \times 10$.

References

1. Blom K, Godec A (2021) Criticality in cell adhesion. Phys Rev X 11(3):031067. [Online]. Available: https://link.aps.org/doi/10.1103/PhysRevX.11.031067
2. Bell G (1978) Models for the specific adhesion of cells to cells Science 200(4342):618–627. [Online]. Available: http://www.sciencemag.org/cgi/doi/10.1126/science.347575

3. Seifert U (2000) Rupture of multiple parallel molecular bonds under dynamic loading. Phys Rev Lett 84(12):2750–2753. [Online]. Available: https://link.aps.org/doi/10.1103/PhysRevLett.84.2750
4. Erdmann T, Schwarz US (2004) Stochastic dynamics of adhesion clusters under shared constant force and with rebinding. J Chem Phys 121(18):8997–9017. [Online]. Available: https://aip.scitation.org/doi/10.1063/1.1805496
5. Dasanna AK, Gompper G, Fedosov DA (2020) Stability of heterogeneous parallel-bond adhesion clusters under load. Phys Rev Res 2:043063. [Online]. Available: https://link.aps.org/doi/10.1103/PhysRevResearch.2.043063
6. Smith A-S, Sackmann E (2009) Progress in mimetic studies of cell adhesion and the mechanosensing. ChemPhysChem 10(1):66–78. [Online]. Available: https://onlinelibrary.wiley.com/doi/abs/10.1002/cphc.200800683
7. Reister-Gottfried E, Sengupta K, Lorz B et al (2008) Dynamics of specific vesicle-substrate adhesion: from local events to global dynamics. Phys Rev Lett 101(20):208103
8. Smith A-S, Sengupta K, Goennenwein S et al (2008) Force-induced growth of adhesion domains is controlled by receptor mobility. Proc Natl Acad Sci USA 105(19):6906–6911, 2008. [Online]. Available: https://www.pnas.org/content/105/19/6906
9. Morone N, Nakada C, Umemura Y et al (2008) Three-dimensional molecular architecture of the plasma-membrane-associated cytoskeleton as reconstructed by freeze-etch electron tomography. In: Introduction to electron microscopy for biologists, series methods in cell biology. Academic Press, vol 88, pp 207–236. [Online]. Available: http://www.sciencedirect.com/science/article/pii/S0091679X08004123
10. Tolentino TP, Wu J, Zarnitsyna VI et al (2008) Measuring diffusion and binding kinetics by contact area FRAP. Biophys J 95(2):920–930. [Online]. Available: http://www.sciencedirect.com/science/article/pii/S0006349508702662
11. Huppa JB, AxmannM, Mörtelmaier MA et al (2010) TCR-peptide-MHC interactions in situ show accelerated kinetics and increased affinity. Nature 463(7283):963–967. [Online]. Available: https://www.nature.com/articles/nature08746
12. Baronsky T, Ruhlandt D, Brückner BR et al (2017) Cell-substrate dynamics of the epithelial-to-mesenchymal transition. Nano Lett 17(5):3320–3326. [Online]. Available: https://doi.org/10.1021/acs.nanolett.7b01558
13. Limozin L, Sengupta K (2009) Quantitative reflection interference contrast microscopy (RICM) in soft matter and cell adhesion. ChemPhysChem 10(16):2752–2768. [Online]. Available: https://onlinelibrary.wiley.com/doi/abs/10.1002/cphc.200900601
14. Fällman E, Schedin S, Jass J et al (2004) Optical tweezers based force measurement system for quantitating binding interactions: system design and application for the study of bacterial adhesion. Biosens Bioelectron 19(11):1429–1437. [Online]. Available: http://www.sciencedirect.com/science/article/pii/S0956566303004615
15. Alon R, Hammer DA, Springer TA (1995) Lifetime of the P-selectin-carbohydrate bond and its response to tensile force in hydrodynamic flow. Nature 374(6522):539–542. [Online]. Available: https://www.nature.com/articles/374539a0
16. Juliano RL, Gagalang E (1977) The adhesion of Chinese hamster cells. I. Effects of temperature, metabolic inhibitors and proteolytic dissection of cell surface macromolecules. J Cell 92(2):209–220. [Online]. Available: https://onlinelibrary.wiley.com/doi/abs/10.1002/jcp.1040920209
17. Piper JW, Swerlick RA, Zhu C (1998) Determining force dependence of two-dimensional receptor-ligand binding affinity by centrifugation. Biophys J 74(1):492–513. [Online]. Available: https://www.ncbi.nlm.nih.gov/pmc/articles/PMC1299402/
18. Marlin SD, Springer TA (1987) Purified intercellular adhesion molecule-1 (ICAM-1) is a ligand for lymphocyte function-associated antigen 1 (LFA-1). Cell 51(5):813–819. [Online]. Available: http://www.sciencedirect.com/science/article/pii/0092867487901048
19. Chen W, Evans EA, McEver RP et al (2008) Monitoring receptor-ligand interactions between surfaces by thermal fluctuations Biophys J 94(2):694–701. [Online]. Available: http://www.sciencedirect.com/science/article/pii/S0006349508707471

20. Evans E, Ritchie K, Merkel R (1995) Sensitive force technique to probe molecular adhesion and structural linkages at biological interfaces. Biophys J 68(6):2580–2587. [Online]. Available: https://www.ncbi.nlm.nih.gov/pmc/articles/PMC1282168/
21. Prechtel K, Bausch AR, Marchi-Artzner V et al (2002) Dynamic force spectroscopy to probe adhesion strength of living cells. Phys Rev Lett 89(2):028101. [Online]. Available: https://link.aps.org/doi/10.1103/PhysRevLett.89.028101
22. Lomakina EB, Waugh RE (2004) Micromechanical tests of adhesion dynamics between neutrophils and immobilized ICAM-1. Biophys J 86(2):1223–1233. [Online]. Available: http://www.sciencedirect.com/science/article/pii/S000634950474196X
23. Evans E, Ritchie K (1997) Dynamic strength of molecular adhesion bonds. Biophys J 72(4):1541–1555. [Online]. Available: https://linkinghub.elsevier.com/retrieve/pii/S0006349597788027
24. Heymann B, Grubmüller H (2000) Dynamic force spectroscopy of molecular adhesion bonds. Phys Rev Lett 84(26):6126–6129. [Online]. Available: https://link.aps.org/doi/10.1103/PhysRevLett.84.6126
25. Sanyour H, Childs J, Meininger GA et al (2018) Spontaneous oscillation in cell adhesion and stiffness measured using atomic force microscopy. Sci Rep. 8(1):1–10. [Online]. Available: https://www.nature.com/articles/s41598-018-21253-9
26. Rico F, Chu C, Abdulreda MH et al (2010) Temperature modulation of integrin-mediated cell adhesion. Biophys J 99(5):1387–1396. [Online]. Available: https://www.ncbi.nlm.nih.gov/pmc/articles/PMC2931747/
27. Sagvolden G, Giaever I, Pettersen EO et al. (1999) Cell adhesion force microscopy. Proc Natl Acad Sci USA 96(2):471–476. [Online]. Available: http://www.pnas.org/cgi/doi/10.1073/pnas.96.2.471
28. Makarov DE (2009) A theoretical model for the mechanical unfolding of repeat proteins. Biophys J 96(6):2160–2167. [Online]. Available: http://www.sciencedirect.com/science/article/pii/S0006349509003208
29. Makarov DE, Wang Z, Thompson JB et al (2002) On the interpretation of force extension curves of single protein molecules. J Chem Phys 116(17):7760–7765. [Online]. Available: https://doi.org/10.1063/1.1466835
30. Williams TE, Nagarajan S, Selvaraj P et al (2001) Quantifying the impact of membrane microtopology on effective two-dimensional affinity. J Biol Chem 276(16):13283–13288. [Online]. Available: http://www.jbc.org/content/276/16/13283
31. Wu Y, Vendome J, Shapiro L et al (2011) Transforming binding affinities from 3D to 2D with application to cadherin clustering. Nature 475(7357):510–513. [Online]. Available: https://www.ncbi.nlm.nih.gov/pmc/articles/PMC3167384/
32. Zarnitsyna V, Zhu C (2012) T cell triggering: insights from 2D kinetics analysis of molecular interactions. Phys Biol 9(4):045005. [Online]. Available: https://doi.org/10.1088%2F1478-3975%2F9%2F4%2F045005
33. Fenz SF, Bihr T, Schmidt D et al (2017) Membrane fluctuations mediate lateral interaction between cadherin bonds. Nat Phys 13(9):906–913. [Online]. Available: https://www.nature.com/articles/nphys4138
34. Merkel R (2001) Force spectroscopy on single passive biomolecules and single biomolecular bonds. Phys Rep 346(5):343–385. [Online]. Available: http://www.sciencedirect.com/science/article/pii/S0370157300001034
35. Gao H, Qian J, Chen B (2011) Probing mechanical principles of focal contacts in cell-matrix adhesion with a coupled stochastic-elastic modelling framework. J R Soc Interface 8(62):1217–1232. [Online]. Available: https://royalsocietypublishing.org/doi/full/10.1098/rsif.2011.0157
36. Hu J, Lipowsky R, Weikl TR (2013) Binding constants of membrane-anchored receptors and ligands depend strongly on the nanoscale roughness of membranes. Proc Natl Acad Sci USA 110(38):15283–15288. [Online]. Available: https://www.pnas.org/content/110/38/15283
37. Steinkühler J, Różycki B, Alvey C et al (2019) Membrane fluctuations and acidosis regulate cooperative binding of 'marker of self' protein cd47 with the macrophage checkpoint receptor sirpα. J Cell Sci 132(4). [Online]. Available: https://doi.org/10.1242/jcs.216770

38. Zuckerman D, Bruinsma R (1995) Statistical mechanics of membrane adhesion by reversible molecular bonds. Phys Rev Lett 74:3900–3903. [Online]. Available: https://link.aps.org/doi/10.1103/PhysRevLett.74.3900
39. Speck T, Reister E, Seifert U (2010) Specific adhesion of membranes: Mapping to an effective bond lattice gas. Phys Rev E 82:021923. [Online]. Available: https://link.aps.org/doi/10.1103/PhysRevE.82.021923
40. Bihr T, Seifert U, Smith A-S (2012) Nucleation of ligand-receptor domains in membrane adhesion. Phys Rev Lett 109(25):258101. [Online]. Available: https://link.aps.org/doi/10.1103/PhysRevLett.109.258101
41. Farago O (2010) Fluctuation-induced attraction between adhesion sites of supported membranes. Phys Rev E 81(5 Pt 1):050902
42. Hu J, Xu G-K, Lipowsky R et al (2015) Binding kinetics of membrane-anchored receptors and ligands: Molecular dynamics simulations and theory. J Chem Phys 143(24):243137. [Online]. Available: https://aip.scitation.org/doi/10.1063/1.4936135
43. Krobath H, Różycki B, Lipowsky R et al (2009) Binding cooperativity of membrane adhesion receptors. Soft Matter 5(17):3354–3361. [Online]. Available: https://pubs.rsc.org/en/content/articlelanding/2009/sm/b902036e
44. Schmidt D, Bihr T, Seifert U et al (2012) Coexistence of dilute and densely packed domains of ligand-receptor bonds in membrane adhesion. EPL 99(3):38003. [Online]. Available: http://stacks.iop.org/0295-5075/99/i=3/a=38003?key=crossref.64617c1745bee09a43069530d47540a6
45. Xu G-K, Hu J, Lipowsky R et al (2015) Binding constants of membrane-anchored receptors and ligands: a general theory corroborated by Monte Carlo simulations. J Chem Phys 143(24):243136. [Online]. Available: http://aip.scitation.org/doi/10.1063/1.4936134
46. Monzel C,Sengupta K (2016) Measuring shape fluctuations in biological membranes. J Phys D 49(24,):243002. [Online]. Available: https://doi.org/10.1088/0022-3727/49/24/243002
47. Gov N, Safran S (2005) Red blood cell membrane fluctuations and shape controlled by atp-induced cytoskeletal defects. Biophys J 88(3):1859–1874. [Online]. Available: https://doi.org/10.1529/biophysj.104.045328
48. Tuvia S, Almagor A, Bitler A et al (1997) Cell membrane fluctuations are regulated by medium macroviscosity: evidence for a metabolic driving force. Proc Natl Acad Sci USA 94(10):5045–5049. [Online]. Available: https://www.ncbi.nlm.nih.gov/pmc/articles/PMC24628/pdf/pq005045.pdf
49. Biswas A, Alex A, Sinha B (2017) Mapping cell membrane fluctuations reveals their active regulation and transient heterogeneities. Biophys J 113:1768ÃÂ¯ÂÂ¿ÂÂ½1781. [Online]. Available: https://www.ncbi.nlm.nih.gov/pmc/articles/PMC5647594/
50. Simson R, Wallraff E, Faix J et al (1998) Membrane bending modulus and adhesion energy of wild-type and mutant cells of dictyostelium lacking talin or cortexillins. Biophys J 74(1):514–522. [Online]. Available: https://www.cell.com/biophysj/fulltext/S0006-3495(98)77808-7
51. Alsteens D, Garcia MC, Lipke PN et al (2010) Force-induced formation and propagation of adhesion nanodomains in living fungal cells. Proc Natl Acad Sci USA 107(48):20744-20749. [Online]. Available: http://dx.doi.org/10.1073/pnas.1013893107
52. Hong Z, Sun Z, Li M et al (2014) Vasoactive agonists exert dynamic and coordinated effects on vascular smooth muscle cell elasticity, cytoskeletal remodelling and adhesion. J Physiol 592(6):1249–1266. [Online]. Available: https://physoc.onlinelibrary.wiley.com/doi/abs/10.1113/jphysiol.2013.264929
53. Zhu Y, Qiu H, Trzeciakowski JP et al (2012) Temporal analysis of vascular smooth muscle cell elasticity and adhesion reveals oscillation waveforms that differ with aging. Aging Cell 11(5):741–750. [Online]. Available: https://onlinelibrary.wiley.com/doi/abs/10.1111/j.1474-9726.2012.00840.x
54. Hong Z, Sun Z, Li Z et al (2012) Coordination of fibronectin adhesion with contraction and relaxation in microvascular smooth muscle. Cardiovasc Res 96(1):73–80. [Online]. Available: https://academic.oup.com/cardiovascres/article-lookup/doi/10.1093/cvr/cvs239

55. Swaminathan V, Mythreye K, O'Brien ET et al (2011) Mechanical stiffness grades metastatic potential in patient tumor cells and in cancer cell lines. Cancer Res 71(15):5075–5080. [Online]. Available: https://cancerres.aacrjournals.org/content/71/15/5075
56. Wang H, Jacobi F, Waschke J et al (2017) Elastic modulus dependence on the specific adhesion of hydrogels. Adv Funct Mater 27(41):1702040. [Online]. Available: https://onlinelibrary.wiley.com/doi/abs/10.1002/adfm.201702040
57. Boudjemaa R, Steenkeste K, Canette A et al (2019) Direct observation of the cell-wall remodeling in adhering Staphylococcus aureus 27217: an AFM study supported by SEM and TEM. Cell Surf 5:100018. [Online]. Available: http://www.sciencedirect.com/science/article/pii/S2468233018300331
58. Huang J, Zarnitsyna VI, Liu B et al (2010) The kinetics of two-dimensional TCR and pMHC interactions determine T-cell responsiveness. Nature 464(7290):932–936. [Online]. Available: https://www.nature.com/articles/nature08944
59. Katsamba P, Carroll K, Ahlsen G et al (2009) Linking molecular affinity and cellular specificity in cadherin-mediated adhesion. Proc Natl Acad Sci USA 106(28):11594–11599. [Online]. Available: https://www.pnas.org/content/106/28/11594
60. Buda A, Pignatelli M (2011) E-cadherin and the cytoskeletal network in colorectal cancer development and metastasis. Cell Commun Adhes 18(6):133–143
61. Korb T, Schlüter K, Enns A et al (2004) Integrity of actin fibers and microtubules influences metastatic tumor cell adhesion. Exp Cell Res 299(1):236–247. [Online]. Available: http://www.sciencedirect.com/science/article/pii/S0014482704003301
62. Zeng Y, Cao Y, Liu L et al (2019) SEPT9_i1 regulates human breast cancer cell motility through cytoskeletal and RhoA/FAK signaling pathway regulation. Cell Death Dis 10(10):1–16. [Online]. Available: https://www.nature.com/articles/s41419-019-1947-9
63. Ising E (1925) Beitrag zur Theorie des Ferromagnetismus. Zeitschrift für Physik 31(1):253–258. [Online]. Available: http://link.springer.com/10.1007/BF02980577
64. Erdmann T, Schwarz US (2004) Stability of adhesion clusters under constant force. Phys. Rev. Lett. 92(10):108102
65. Qian J, Wang J, Gao H (2008) Lifetime and strength of adhesive molecular bond clusters between elastic media†. Langmuir 24(4):1262-1270. [Online]. Available: http://dx.doi.org/10.1021/la702401b
66. Glauber RJ (1963) Time-dependent statistics of the Ising model. J Math Phys 4(2):294–307. [Online]. Available: https://aip.scitation.org/doi/10.1063/1.1703954
67. Kawasaki K (1966) Diffusion constants near the critical point for time-dependent ising models. I. Phys Rev 145:224–230. [Online]. Available: https://link.aps.org/doi/10.1103/PhysRev.145.224
68. Iosifescu M (2014) Finite Markov processes and their applications. Courier Corporation
69. Fowler RH (1939) Statistical thermodynamics. CUP Archive
70. Godec A, Metzler R (2016) Universal proximity effect in target search kinetics in the few-encounter limit. Phys Rev X 6:041037. [Online]. Available: https://link.aps.org/doi/10.1103/PhysRevX.6.041037
71. Hartich D, Godec A (2018) Duality between relaxation and first passage in reversible markov dynamics: rugged energy landscapes disentangled. New J Phys 20(11):112002. [Online]. Available: http://dx.doi.org/10.1088/1367-2630/aaf038
72. Hartich D, Godec A (2019) Interlacing relaxation and first-passage phenomena in reversible discrete and continuous space markovian dynamics. J Stat Mech Theory Exp 2019(2):024002. [Online]. Available: http://dx.doi.org/10.1088/1742-5468/ab00df
73. Onsager L (1944) Crystal statistics. i. a two-dimensional model with an order-disorder transition. Phys Rev 65:117–149. [Online]. Available: https://link.aps.org/doi/10.1103/PhysRev.65.117
74. Ayee MAA, LeMaster E, Teng T et al (2018) Hypotonic challenge of endothelial cells increases membrane stiffness with no effect on tether force. Biophys J 114(4):929–938. [Online]. Available: https://www.ncbi.nlm.nih.gov/pmc/articles/PMC5984970/

75. Callies C, Fels J, Liashkovich I et al (2011) Membrane potential depolarization decreases the stiffness of vascular endothelial cells. J Cell Sci 124(11):1936–1942. [Online]. Available: https://jcs.biologists.org/content/124/11/1936
76. Sliogeryte K, Botto L, Lee DA et al (2016) Chondrocyte dedifferentiation increases cell stiffness by strengthening membrane-actin adhesion. Osteoarthritis Cartilage 24(5):912–920. [Online]. Available: http://www.sciencedirect.com/science/article/pii/S1063458415014296
77. Sanyour HJ, Li N, Rickel AP et al (2019) Membrane cholesterol and substrate stiffness co-ordinate to induce the remodelling of the cytoskeleton and the alteration in the biomechanics of vascular smooth muscle cells. Cardiovasc Res 115(8):1369–1380
78. Fowler PW, Hélie J, Duncan A et al (2016) Membrane stiffness is modified by integral membrane proteins. Soft Matter 12(37):7792–7803. [Online]. Available: https://pubs.rsc.org/en/content/articlelanding/2016/sm/c6sm01186a
79. Dimova R (2014) Recent developments in the field of bending rigidity measurements on membranes. Adv Colloid Interface Sci 208:225–234. [Online]. Available: https://www.sciencedirect.com/science/article/pii/S0001868614001031
80. Faizi HA, Frey SL, Steinkühler J et al (2019) Bending rigidity of charged lipid bilayer membranes. Soft Matter 15(29):6006–6013. [Online]. Available: https://doi.org/10.1039/C9SM00772E
81. Faris ME, Lacoste D, Pécréaux J et al (2009) Membrane tension lowering induced by protein activity. Phys Rev Lett 102(3):038102. [Online]. Available: https://link.aps.org/doi/10.1103/PhysRevLett.102.038102
82. Paszek MJ, Boettiger D, Weaver VM et al (2009) Integrin clustering is driven by mechanical resistance from the glycocalyx and the substrate. PLoS Comput Biol 5(12):e1000604. [Online]. Available: https://doi.org/10.1371/journal.pcbi.1000604
83. Caputo KE, Hammer DA (2005) Effect of microvillus deformability on leukocyte adhesion explored using adhesive dynamics simulations. Biophys J 89(1):187–200. [Online]. Available: https://doi.org/10.1529/biophysj.104.054171
84. Nermut M, Green N, Eason P et al (1988) Electron microscopy and structural model of human fibronectin receptor. The EMBO J 7(13):4093–4099. [Online]. Available: https://doi.org/10.1002/j.1460-2075.1988.tb03303.x
85. Pelta J, Berry H, Fadda G et al (2000) Statistical conformation of human plasma fibronectin Biochemistry 39(17):5146–5154. [Online]. Available: https://doi.org/10.1021/bi992770x
86. Braig S, Schmidt BS, Stoiber K et al (2015) Pharmacological targeting of membrane rigidity: implications on cancer cell migration and invasion. New J Phys 17(8):083007. [Online]. Available: https://doi.org/10.1088/1367-2630/17/8/083007
87. Roli A, Villani M, Filisetti A et al (2018) Dynamical criticality: overview and open questions. J Syst Sci Complexity 31(3):647–663. [Online]. Available: https://doi.org/10.1007/s11424-017-6117-5
88. Doroudian G, Curtis MW, Gang A et al (2013) Cyclic strain dominates over microtopography in regulating cytoskeletal and focal adhesion remodeling of human mesenchymal stem cells. Biochem Biophys Res Commun 430(3):1040–1046. [Online]. Available: http://www.sciencedirect.com/science/article/pii/S0006291X12023728
89. Greiner AM, Chen H, Spatz JP et al (2013) Cyclic tensile strain controls cell shape and directs actin stress fiber formation and focal adhesion alignment in spreading cells. PLoS One 8(10):1–10. [Online]. Available: https://doi.org/10.1371/journal.pone.0077328
90. Parsons JT, Horwitz AR, Schwartz MA (2010) Cell adhesion: integrating cytoskeletal dynamics and cellular tension. Nat Rev Mol Cell Biol 11(9):633–643. [Online]. Available: http://dx.doi.org/10.1038/nrm2957
91. Chesla SE, Selvaraj P, Zhu C (1998) Measuring two-dimensional receptor-ligand binding kinetics by micropipette. Biophys J 75(3):1553–1572. [Online]. Available: https://doi.org/10.1016/S0006-3495(98)74074-3
92. Zhelev DV, Needham D, Hochmuth RM (1994) Role of the membrane cortex in neutrophil deformation in small pipets. Biophys J 67(2):696–705. [Online]. Available: https://www.cell.com/biophysj/pdf/S0006-3495(94)80529-6.pdf

93. Sharma K (1993) Cellular deformability studies in leukemia. Physiol Chem Phys Med NMR 25(4):293–297. [Online]. Available: https://pubmed.ncbi.nlm.nih.gov/8153154/
94. Hasan N, Adams G, Joiner M et al (1998) Hypoxia facilitates tumour cell detachment by reducing expression of surface adhesion molecules and adhesion to extracellular matrices without loss of cell viability. Br J Cancer 77(11):1799–1805. [Online]. Available: http://dx.doi.org/10.1038/bjc.1998.299
95. Yamada KM, Sixt M (2019) Mechanisms of 3d cell migration. Nat Rev Mol Cell Biol 20(12):738-752. [Online]. Available: http://dx.doi.org/10.1038/s41580-019-0172-9
96. De Pascalis C, Etienne-Manneville S (2017) Single and collective cell migration: the mechanics of adhesions. Mol Biol Cell 28(14):1833–1846. [Online]. Available: http://dx.doi.org/10.1091/mbc.e17-03-0134
97. Kirfel G (2004) Cell migration: mechanisms of rear detachment and the formation of migration tracks. Eur J Cell Biol 83(11–12):717–724. [Online]. Available: http://dx.doi.org/10.1078/0171-9335-00421
98. Dwir O, Solomon A, Mangan S et al (2003) Avidity enhancement of l-selectin bonds by flow. J Cell Biol 163(3):649–659. [Online]. Available: http://dx.doi.org/10.1083/jcb.200303134
99. Erbeldinger N, Rapp F, Ktitareva S et al (2017) Measuring leukocyte adhesion to (primary) endothelial cells after photon and charged particle exposure with a dedicated laminar flow chamber. Front Immunol 8:627. [Online]. Available: https://www.frontiersin.org/article/10.3389/fimmu.2017.00627
100. Zhou Y, Kucik DF, Szalai AJ et al (2014) Human neutrophil flow chamber adhesion assay. JoVE 89:e51410. [Online]. Available: https://www.ncbi.nlm.nih.gov/pmc/articles/PMC4166108/
101. Cross SE, Jin Y-S, Rao J et al (2007) Nanomechanical analysis of cells from cancer patients. Nat Nanotechnol 2(12):780–783. [Online]. Available: http://dx.doi.org/10.1038/nnano.2007.388
102. Plischke M, Mattis D (1970) Two-dimensional ising model in a finite magnetic field. Phys Rev B 2:2660–2663. [Online]. Available: https://link.aps.org/doi/10.1103/PhysRevB.2.2660
103. Morgenstern I, Binder K, Hornreich RM (1981) Two-dimensional ising model in random magnetic fields. Phys Rev B 23:287–297. [Online]. Available: https://link.aps.org/doi/10.1103/PhysRevB.23.287
104. Stump DR (1987) Entropy of the two-dimensional ising model. Phys Rev A 36:4439–4446. [Online]. Available: https://link.aps.org/doi/10.1103/PhysRevA.36.4439
105. Kaufman M (1987) Square-lattice ising model in a weak uniform magnetic field: renormalization-group analysis. Phys Rev B 36:3697–3700. [Online]. Available: https://link.aps.org/doi/10.1103/PhysRevB.36.3697
106. de With G (2013) Liquid-state physical chemistry: fundamentals, modeling, and applications. Wiley-VCH Verlag
107. Datta R, Acharyya M, Dhar A (2018) Magnetisation reversal in ising ferromagnet by thermal and field gradients. Heliyon 4(10):e00892. [Online]. Available: https://www.sciencedirect.com/science/article/pii/S2405844018314142
108. Brendel K, Barkema GT, van Beijeren H (2003) Magnetization reversal times in the two-dimensional ising model. Phys Rev E 67:026119. [Online]. Available: https://link.aps.org/doi/10.1103/PhysRevE.67.026119
109. García-Pablos D, García-Mochales P, Garcia N et al (1996) Nonhomogeneous magnetization reversal in 2d ising clusters. J Appl Phys 79(8):6019–6021. [Online]. Available: https://aip.scitation.org/doi/abs/10.1063/1.362139
110. Brendel K, Barkema G, van Beijeren H (2005) Nucleation time distribution in the two-dimensional ising model with spin-flip dynamics. AIP Conf Proc 800(1):39–49. [Online]. Available: https://aip.scitation.org/doi/abs/10.1063/1.2138592
111. Sear RP (2006) Heterogeneous and homogeneous nucleation compared: Rapid nucleation on microscopic impurities. J Phys Chem B 110(10):4985–4989. [Online]. Available: https://pubs.acs.org/doi/10.1021/jp056377e

112. Hohenberg PC, Halperin BI (1977) Theory of dynamic critical phenomena. Rev Mod Phys 49:435–479. [Online]. Available: https://link.aps.org/doi/10.1103/RevModPhys.49.435
113. Binder K, Young AP (1986) Spin glasses: Experimental facts, theoretical concepts, and open questions. Rev Mod Phys 58:801–976. [Online]. Available: https://link.aps.org/doi/10.1103/RevModPhys.58.801
114. Krobath H, Różycki B, Lipowsky R et al (2011) Line tension and stability of domains in cell-adhesion zones mediated by long and short receptor-ligand complexes. PLoS One 6(8):e23284. [Online]. Available: https://dx.plos.org/10.1371/journal.pone.0023284
115. Erdmann T, Schwarz US (2007) Impact of receptor-ligand distance on adhesion cluster stability. Eur Phys J E 22(2):123–137. [Online]. Available: https://doi.org/10.1140/epje/e2007-00019-8
116. Muñoz MA(2018) Colloquium: criticality and dynamical scaling in living systems. Rev Mod Phys 90:031001. [Online]. Available: https://link.aps.org/doi/10.1103/RevModPhys.90.031001
117. Balaban NQ, Schwarz US, Riveline D et al (2001) Force and focal adhesion assembly: a close relationship studied using elastic micropatterned substrates. Nat Cell Biol 3(5):466–472. [Online]. Available: http://dx.doi.org/10.1038/35074532
118. Thomas WE, Vogel V, Sokurenko E (2008) Biophysics of catch bonds. Annu Rev Biophys 37(1):399–416. [Online]. Available: https://doi.org/10.1146/annurev.biophys.37.032807.125804
119. Chakrabarti S, Hinczewski M, Thirumalai D (2014) Plasticity of hydrogen bond networks regulates mechanochemistry of cell adhesion complexes. Proc Natl Acad Sci USA 111(25):9048–9053. [Online]. Available: https://www.pnas.org/content/111/25/9048
120. Chakrabarti S, Hinczewski M, Thirumalai D (2017) Phenomenological and microscopic theories for catch bonds. J Struct Biol 197(1):50–56. [Online]. Available: https://www.sciencedirect.com/science/article/pii/S1047847716300582
121. Adhikari S, Moran J, Weddle C et al (2018) Unraveling the mechanism of the cadherin-catenin-actin catch bond. PLoS Comput Biol 14(8):1–17. [Online]. Available: https://doi.org/10.1371/journal.pcbi.1006399
122. Zhuravlev PI, Hinczewski M, Chakrabarti S et al (2016) Force-dependent switch in protein unfolding pathways and transition-state movements. Proc Natl Acad Sci USA 113(6):E715–E724. [Online]. Available: https://www.pnas.org/content/113/6/E715
123. Blom K, Godec A (2021) https://gitlab.gwdg.de/kblom/criticality-in-cell-adhesion
124. Mulla Y, Oliveri G, Overvelde JTB et al (2018) Crack initiation in viscoelastic materials. Phys Rev Lett 120:268002. [Online]. Available: https://link.aps.org/doi/10.1103/PhysRevLett.120.268002
125. Bender CM, Orszag SA (2013) Advanced mathematical methods for scientists and engineers I: asymptotic methods and perturbation theory. Springer Science & Business Media
126. Cahill ND, D'Errico JR, Narayan DA et al (2002) Fibonacci determinants. College Math J 33(3):221–225. [Online]. Available: https://doi.org/10.1080/07468342.2002.11921945
127. Jia J-T (2018) Numerical algorithms for the determinant evaluation of general hessenberg matrices. J Math Chem 56(1):247–256. [Online]. Available: https://doi.org/10.1007/s10910-017-0794-0
128. Hartich D, Godec A (2019) Extreme value statistics of ergodic markov processes from first passage times in the large deviation limit. J Phys A Math Theor 52(24):244001. [Online]. Available: https://doi.org/10.1088%2F1751-8121%2Fab1eca
129. Hänggi P, Talkner P, Borkovec M (1990) Reaction-rate theory: fifty years after Kramers. Rev Mod Phys 62:251–341. [Online]. Available: https://link.aps.org/doi/10.1103/RevModPhys.62.251
130. Wannier GH (1945) The statistical problem in cooperative phenomena. Rev Mod Phys 17:50–60. [Online]. Available: https://link.aps.org/doi/10.1103/RevModPhys.17.50

Chapter 6
Global Speed Limit for Finite-Time Dynamical Phase Transition in Nonequilibrium Relaxation

I learned very early the difference between knowing the name of something, and knowing something.

Richard Feynmann in The Pleasure Of Finding Things Out

Recent works [1–4] unraveled an intriguing **finite-time dynamical phase transition** in the thermal relaxation of the MF Curie-Weiss model. The phase transition reflects a sudden switch in the magnetization dynamics. Its existence in systems with a finite range of interaction, however, remained unclear. Employing the BG approximation, which is exact on Bethe lattices, we here demonstrate the finite-time dynamical phase transition in nearest-neighbor Ising systems for arbitrary quenches, including those within the two-phase region. Strikingly, for any given initial condition we prove and explain the existence of non-trivial **speed limits** for the dynamical phase transition and the relaxation of magnetization, which are absent in the MF setting. Pair correlations, which are neglected in MF theory and trivial in the Curie-Weiss model, account for kinetic constraints due to frustrated local configurations that give rise to a global speed limit. This chapter is based on the work presented in [5].

6.1 Introduction and Motivation

Despite its overwhelming importance in condensed matter physics [6, 7], our understanding of thermal relaxation kinetics is far from complete, and mostly limited to systems near equilibrium [8–10] and non-equilibrium [11–13] steady states. Notable advances in understanding relaxation dynamics out of equilibrium include far-from-equilibrium fluctuation-dissipation theorems [14, 15], “frenesy” [16], anomalous relaxation a.k.a. the Mpemba effect [17–20], optimal heating and cooling [21] as

K. Blom, *Pair-Correlation Effects in Many-Body Systems*, Springer Theses,
https://doi.org/10.1007/978-3-031-29612-3_6

well as driving [22, 23] protocols, asymmetries in heating and cooling rates [24–27], and dynamical phase transitions (i.e. the occurence of non-analytic points in distributions of physical observables) [28–52]. Further important results on non-equilibrium relaxation are embodied in thermodynamic uncertainty relations for non-stationary systems [53–58], and so called *speed limits* [59–62, 62–75, 75–80].

In contrast to the well established concept of quantum speed limits [59–62, 62–70] that has long been known [59], it was comparably only recently found that the evolution of classical systems is also bounded by fundamental speed limits [71–77]. Quantum and classical speed-limits impose an upper bound on the rate of change of a system state evolving from a given non-stationary initial state, and arise as an intrinsic dynamical property of Hilbert space [71]. Moreover, it was found that by considering the thermodynamic cost of the state change, one may derive even sharper thermodynamic speed limits that bound the rate of change of a system state from above by the entropy production rate [73, 75, 78–80].

Recently, a surprising *finite-time dynamical phase transition* was observed in a MF Ising system [1, 2], manifested as a **finite-time singularity** [3, 4] in the probability density of magnetization [1] and entropy flow per spin [2] upon a quench from any sub-critical temperature $T < T_c$ to a temperature T_q[1]. In contrast to conventional phase transitions, here time plays the role of a control parameter inducing an abrupt change of the typical dynamics [1, 2]. The sudden transition from a **Gibbsian** to a **non-Gibbsian** probability density occurs for *all* quenches from sub-critical temperatures $T < T_c$, whereby the initial location of the singularity depends on T and T_q [4]. Upon quenches from super- critical temperatures $T > T_c$, the probability density remains Gibbsian forever [4], but the dynamics is non-ergodic [81].

Notwithstanding the detailed results on the non-Gibbsian transition in the MF setting, it remains unknown if and in what form this dynamical phase transition exists in systems with a finite range of interactions. Moreover, since speed limits bound from below the time of reaching a final state from a given initial state, the following intriguing questions arise:

> What happens with the speed limit in the finite-time dynamical phase transition, where the dynamics undergoes an abrupt change? Is there a global speed limit to reach the critical time?

To shed light on these questions, we here present analytical results on non-equilibrium relaxation of nearest-neighbor Ising systems on the BG level [82, 83]. Our results confirm, for the first time, the existence of the finite-time dynamical phase transition in finite-range Ising systems. Strikingly, we derive *explicit* global speed limits to both, the **critical time** and relaxation time, which are absent in the MF setting. Notably, the speed limit is saturated by an antiferromagnetic interaction and is faster than the dynamics of a non-interacting system. Accounting for unfavorable local spin configurations, pair correlations, which are neglected in MF theory, impose a global speed limit in the dynamical phase transition.

[1] In [1, 2] only super-critical quench temperatures $T_q > T_c$ are considered, whereas [3, 4] consider all possible T_q above the initial temperature T.

6.2 Fundamentals

In previous chapters we defined the free energy density $\tilde{\mathrm{f}}$ in terms of the fraction of down spins φ, see e.g. Eq. (2.14). However, to be consistent with the literature [1, 2] addressed in this chapter, we will define $\tilde{\mathrm{f}}$ in terms of the magnetization per spin $m(\boldsymbol{\sigma}) \equiv N^{-1}\sum_{i=1}^{N}\sigma_i$, which is related to φ through the simple relation $\varphi = (1-m)/2$.

6.2.1 Equilibrium

We consider an Ising model in the absence of a magnetic field. The Hamiltonian $\mathcal{H}(\boldsymbol{\sigma})$ is given by Eq. (1.1) with $h = 0$, and J denotes the ferromagnetic ($J > 0$) or antiferromagnetic ($J < 0$) coupling. The spins are placed on a Bethe lattice with coordination number $\bar{z} \in \mathbb{N}^+$. Figure 6.1a shows three examples of Bethe lattices. The equilibrium free energy density (in units of $k_\mathrm{B}T$, $\tilde{J} \equiv J/k_\mathrm{B}T$) in the thermodynamic limit is defined as $\tilde{\mathrm{f}}(m, \tilde{J}) = \lim_{N\to\infty}^{m=\text{const.}}\left[N^{-1}\ln(\mathcal{Z}_m)\right]$,[2] where $\mathcal{Z}_k \equiv \sum_{\boldsymbol{\sigma}}\exp(-\mathcal{H}(\boldsymbol{\sigma})/k_\mathrm{B}T)\delta_{m(\boldsymbol{\sigma}),k}$ is the fixed-magnetization partition function (see Sect. 1.1.5) with Kronecker delta $\delta_{a,b}$ being 1 when $a = b$ and 0 otherwise. Within BG theory, the free energy density, $\tilde{\mathrm{f}}_\mathrm{BG}(m, \tilde{J}) = \tilde{\mathrm{u}}_\mathrm{BG}(m, \tilde{J}) - \tilde{\mathrm{s}}_\mathrm{BG}(m, \tilde{J})$, is given by Eq. (2.14) with $m = 1 - 2\varphi$. Similarly, the MF free energy density, $\tilde{\mathrm{f}}_\mathrm{MF}(m, \tilde{J})$, is given by Eq. (2.40). The BG critical coupling above which $\tilde{\mathrm{f}}_\mathrm{BG}(m, \tilde{J})$ develops two minima is given by $\tilde{J}_\mathrm{crit}^\mathrm{BG} \equiv \ln(\bar{z}/[\bar{z}-2])/2$ (see Sect. 2.8.3), and correctly diverges in dimension one with $\bar{z} = 2$, where no phase transition occurs.

6.2.2 Kinetics

We now introduce **continuous-time stochastic dynamics** for changes of the magnetization within the local equilibrium approximation (LEQ) approximation [84, 85] (see also Sect. 5.4), which is highly accurate in the thermodynamic limit [86]. Let $W^\pm(M, \tilde{J})$ denote the transition rate to change the total magnetization from $M \equiv Nm \to M \pm 2$ by a single-spin flip. Following [84, 85], we define, in the thermodynamic limit, an intensive transition rate $w^\pm(m, \tilde{J}) \equiv \lim_{N\to\infty}^{m=\text{const.}}[W^\pm(Nm, \tilde{J})/N]$. Let τ be an intrinsic time-scale of infinitesimal changes of magnetization $m \to m + \mathrm{d}m$ [87], and $\zeta_\mathrm{BG}^\dagger(1/2 - m/2)$ the fraction of defects for a given magnetization m given by Eq. (2.12). Then, the BG-LEQ transition rate reads

$$w_\mathrm{BG}^\pm(m, \tilde{J}) = \frac{1 \mp m}{2\tau}\left(\mathrm{e}^{-\tilde{J}} + \frac{2\zeta_\mathrm{BG}^\dagger(1/2 - m/2)\sinh(\tilde{J})}{1 \mp m}\right)^{\bar{z}}. \tag{6.1}$$

[2] In this chapter we write $\tilde{\mathrm{f}}(m, \tilde{J})$ instead of $\tilde{\mathrm{f}}(m)$, because $\tilde{J}$ will be considered a variable instead of a parameter. The free energy densities defined in Eqs. (2.14) and (2.40) remain the same.

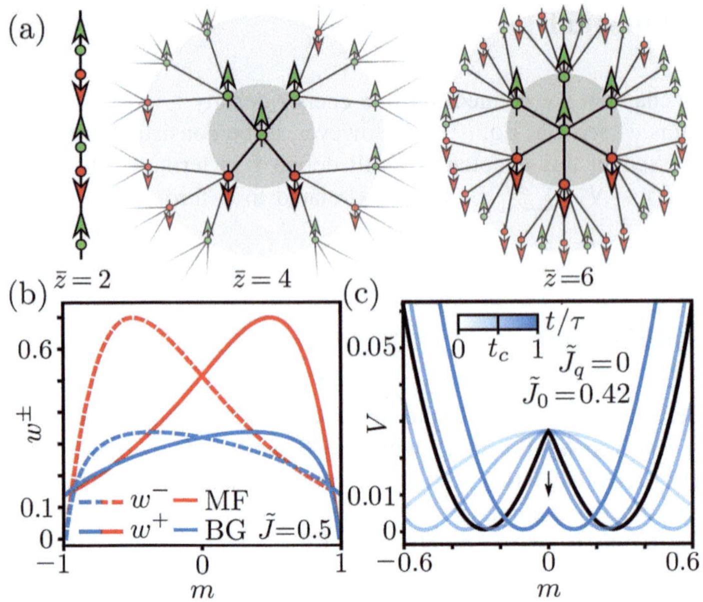

Fig. 6.1 Finite-time dynamical phase transition on the Bethe lattice. a Examples of Bethe lattices. In **b, c** we consider a Bethe lattice with $\bar{z} = 4$. **b** Forward (solid lines) and backward (dashed lines) rates $w^{\pm}$ within the BG [blue; Eq. (6.1)] and MF approximation (red; see [1]). **c** Temporal evolution of $V(m, \tilde{J}, t)$ for a quench into the one-phase domain. Time passes from light to dark blue lines. At the critical time t_c (black line) a cusp emerges at $m = 0$

The transition rates obey the **parity symmetry** $w^{\pm}_{\text{BG}}(m, \tilde{J}) = w^{\mp}_{\text{BG}}(-m, \tilde{J})$ and DB w.r.t. the free energy density, $w^{+}_{\text{BG}}(m, \tilde{J})/w^{-}_{\text{BG}}(m, \tilde{J}) = \exp(-2\partial_m \tilde{\mathrm{f}}_{\text{BG}}(m, \tilde{J}))$. In the weak coupling (or high temperature) limit we recover the MF transition rates $\lim_{\tilde{J}\to 0} w^{\pm}_{\text{BG}}(m, \tilde{J}) = w^{\pm}_{\text{MF}}(m, \tilde{J}) + \mathcal{O}(\tilde{J}^2)$ reported in [1] (see Eq. (3) therein). A comparison between the BG and MF transition rates for $(\bar{z}, \tilde{J}) = (4, 0.5)$ is shown in Fig. 6.1b.

6.2.3 Kinetics in the Thermodynamic Limit

Let $P_N(m, \tilde{J}, t)$ be the probability density of m at time t evolving according to the incoming and outgoing local fluxes $\partial_t P_N(m, \tilde{J}, t) = j^+(m, t) - j^-(m, t)$. The time-dependent **large-deviation rate function** is defined as

$$V(m, \tilde{J}, t) \equiv -\lim_{N\to\infty} N^{-1} \ln P_N(m, \tilde{J}, t).$$

At equilibrium the rate function is given by the free energy density $V_{\text{eq}}(m, \tilde{J}) \equiv \tilde{\mathrm{f}}(m, \tilde{J}) - \tilde{\mathrm{f}}(\bar{m}, \tilde{J})$, with $\bar{m}(\tilde{J}) \equiv \arg\min_m \tilde{\mathrm{f}}(m, \tilde{J})$ denoting the location of free energy minima. Out of equilibrium $V(m, \tilde{J}, t)$ obeys a **Hamilton-Jacobi**

equation (HJ) [1, 88, 89]

$$\partial_t V(m, \tilde{J}, t) + \mathcal{H}(m, \partial_m V(m, \tilde{J}, t)) = 0, \tag{6.2}$$

with the Hamiltonian given by

$$\mathcal{H}(q, p) = w^+(q, \tilde{J})(e^{2p} - 1) + w^-(q, \tilde{J})(e^{-2p} - 1). \tag{6.3}$$

Equation (6.2) can be derived directly from the master equation for $P_N(m, \tilde{J}, t)$ as the instanton solution in the thermodynamic limit. We are interested in the evolution of $V(m, \tilde{J}, t)$ upon a quench $\tilde{J}_0 \to \tilde{J}_q < \tilde{J}_0$, where $\tilde{J}_q$ may be positive or negative. For a quench towards a lower coupling strength (or higher temperature), the spins undergo a **disordering transition** and break the ferromagnetic state. Experimentally quenches to negative $\tilde{J}_q$ may be achieved, e.g. by ultrafast optical switching ferro-antiferromagnetic materials [90] or by spin-population inversion in metals by radio-frequency irradiation [91] yielding negative spin temperatures.[3] Note that quenches beyond the **Néel point** (i.e. the antiferromagnetic critical coupling) push the system across the antiferromagnetic transition, which m does not detect [93–96]. In fact, quenching from the antiferromagnetic two-phase region and replacing m with the **staggered magnetization** [93–96] yields mirror-symmetric results (see Appendix 7).

6.3 Dynamical Phase Transition

We throughout assume that the system is initially prepared at equilibrium in the two-phase regime $\tilde{J}_0 > \tilde{J}^{\mathrm{BG}}_{\mathrm{crit}}$ (i.e. above the critical coupling), and thus $V_{\mathrm{BG}}(m, \tilde{J}, 0) = \tilde{\mathrm{f}}_{\mathrm{BG}}(m, \tilde{J}_0) - \tilde{\mathrm{f}}_{\mathrm{BG}}(\bar{m}, \tilde{J}_0)$. At $t = 0$ we apply an instantaneous quench $\tilde{J}_q < \tilde{J}_0$ by changing the temperature T or the coupling J, which pushes the system out of equilibrium. The rate function $V_{\mathrm{BG}}(m, \tilde{J}, t > 0)$ thereupon evolves according to Eq. (6.2), which we solve numerically (see Fig. 6.1d and 6.2a for quenches with $\tilde{J}_q \le \tilde{J}^{\mathrm{BG}}_{\mathrm{crit}}$ and $\tilde{J}_q \ge \tilde{J}^{\mathrm{BG}}_{\mathrm{crit}}$, respectively). As $V_{\mathrm{BG}}(m, \tilde{J}, t)$ relaxes towards the new equilibrium at $\tilde{J}_q$, there is a defined moment $t_c^{\mathrm{BG}}(\tilde{J}_0, \tilde{J}_q)$—the **critical time**—where $V_{\mathrm{BG}}(m, \tilde{J}, t)$ *abruptly* develops a **cusp** (black line in Figs. 6.1c and 6.2a) and becomes non-Gibbsian. The phenomenon was coined finite-time dynamical phase transition [1, 3, 4] and is hereby confirmed in nearest-neighbor Ising systems.

Due to the reflection symmetry around $m = 0$, and local rates w^+_{BG} and w^-_{BG} that are strictly increasing and decreasing, respectively, in an interval around $m = 0$ (see Fig. 6.1b), the forward and backward probability fluxes, $j^{\pm}(m, t)$, remain perfectly balanced in a region around $m = 0$ during a transient period after the quench (see

[3] Note that only the nuclear spin temperature becomes negative, other degrees of freedom actually heat up. For an excellent pedagogical expose on negative temperatures in systems with bounded energy spectra see [92].

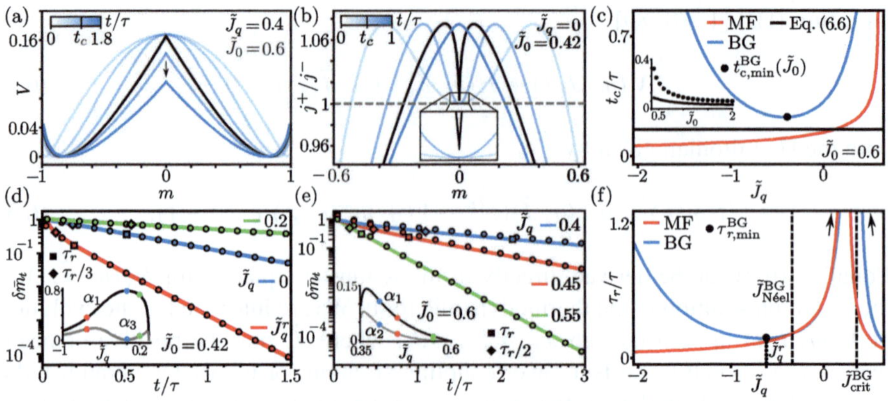

Fig. 6.2 Finite-time dynamical phase transition for quenches in the two-phase domain, scaling of the critical time, and relaxation dynamics of the rate function minima. **a** Temporal evolution of the BG rate function $V(m, \tilde{J}_q, t)$ upon a quench into the two-phase domain. **b** Temporal evolution of the probability-flux ratio $j^+(m,t)/j^-(m,t)$ after a quench into the one-phase regime. At the critical time t_c (black line) the ratio discontinuously jumps to a value above 1 at $m = 0$. Inset: Enlargement around $m = 0$. **c** BG (blue) and MF (red) critical time t_c/τ as a function of $\tilde{J}_q$. The BG critical time attains a global minimum $t^{\mathrm{BG}}_{c,\min}$ (black dot) for an antiferromagnetic quench, bounded from below by Eq. (6.6) (black line); Inset: $t^{\mathrm{BG}}_{c,\min}$ (black dots) and Eq. (6.6) (black line) as a function of $\tilde{J}_0$. **d, e** Relaxation of excess mean magnetization $\delta\bar{m}_t = (\bar{m}_t - \bar{m}_\infty)/(\bar{m}_0 - \bar{m}_\infty)$ upon a quench in the one- (**d**) and two-phase (**e**) domain. Dots depict the first two nonzero terms of the analytical power series solution, lines are numerical solutions of the differential equation. Squares/diamonds denote the first τ_r and second ($\tau_r/3$ in **d** and $\tau_r/2$ in **e**) relaxation time-scales, respectively. Inset: First two nonzero prefactors of the power series which enter Eq. (6.7). (**f**) BG (blue) and MF (red) relaxation time τ_r/τ as a function of $\tilde{J}_q$. τ_r^{BG} has a local minimum at $\tilde{J}_q^r < 0$ [see Eq. (6.8)]. In all panels $\bar{z} = 4$

Fig. 6.2b). As a result, $P_N(m \approx 0, \tilde{J}, t)$ is transiently "locked" in the initial state (see Fig. 6.1c and Fig. 6.2a). "Fronts" of net flux towards $m = 0$ gradually develop on each side and drift towards the center (Fig. 6.2b). Once the fronts collide, the dynamical phase transition takes place as an instability, in which the flux ratio $j^+(0,t)/j^-(0,t)$ discontinuously jumps to a value larger than 1 (inset of 6.2b). At the transition the dynamics switches from *confined in the wells* to *exploring the free energy barrier*, i.e. between the formation of defects in ordered domains to their (partial) melting.

The fact that the cusp appears upon quenches within the two-phase regime, $\tilde{J}^{\mathrm{BG}}_{\mathrm{crit}} \leq \tilde{J}_q < \tilde{J}_0$ (see Fig. 6.2a), implies that the dynamical phase transition does not require a change in geometry from a double- to a single-well potential. Moreover, we show in Appendix 3.2 that the initial location of the cusp undergoes a symmetry-breaking transition above a threshold coupling $\tilde{J}_0 > \tilde{J}_0^{\mathrm{SB}}(\tilde{J}_q)$ whereupon it moves from the center $m = 0$. For infinite temperature/zero coupling quenches the symmetry-breaking transition occurs at $\tilde{J}^{\mathrm{SB}}_{0,\mathrm{BG}}(0) = \ln([\bar{z}+1]/[\bar{z}-2])/2$, which in the MF setting simplifies to $\lim_{\bar{z}\to\infty} \tilde{J}^{\mathrm{SB}}_{0,\mathrm{BG}}(0) = 3/(2\bar{z}) + \mathcal{O}(1/\bar{z}^2)$ and agrees with the results in [3, 4].

6.4 Critical Time

We now determine the critical time t_c, i.e. the first instance a cusp appears at $m = 0$. The critical time can be determined from the curvature [1] or slope [3, 4] at $m = 0$ and reads (see derivation in Appendix 3)

$$t_c(\tilde{J}_0, \tilde{J}_q) = \frac{\ln(1-\tilde{f}''(0, \tilde{J}_q)/\tilde{f}''(0, \tilde{J}_0))}{8w^{\pm}(0, \tilde{J}_q)\tilde{f}''(0, \tilde{J}_q)}, \tag{6.4}$$

where $\tilde{f}''(0, \tilde{J}) \equiv d^2\tilde{f}(m, \tilde{J})/dm^2|_{m=0}$ is the curvature of the barrier. Inserting the MF free energy density [see Eq. (2.40)] and transition rates (see Eq. (3) in [1]) into Eq. (6.4), we recover the results derived in [1, 3, 4]. For the BG critical time we insert Eqs. (2.14) and (6.1), respectively. In Fig. 6.2c the BG (blue) and MF (red) critical times are shown as a function of $\tilde{J}_q$ for $(\bar{z}, \tilde{J}_0) = (4, 0.6)$, and display starkly dissimilar behavior. In particular, the BG critical time displays a global minimum—a global speed limit—that is absent in the MF setting. This implies a dominant role of local spin configurations, which are accounted for in the BG theory but ignored in MF theory.

6.5 Lower Bounds on the Critical Time

The stationary points of Eq. (6.4) cannot be determined analytically. To confirm that the speed limit indeed exists, we instead prove lower bounds on Eq. (6.4) for quenches in the one- and two-phase domain.

6.5.1 *Lower Bound for Quenches in the Two-Phase Domain*

The critical time $t_c(\tilde{J}_0, \tilde{J}_q)$ is monotonically increasing with $\tilde{J}_q$ for $\tilde{J}_{\text{crit}}^{\text{BG}} \leq \tilde{J}_q < \tilde{J}_0$ (see proof in Appendix 4.2). Thus, the critical time for quenches within the two-phase regime is bounded from below by $t_c^{\text{BG}}(\tilde{J}_0, \tilde{J}_{\text{crit}}^{\text{BG}})$, which gives the following $\tilde{J}_q$-independent lower bound

$$t_c^{\text{BG}}(\tilde{J}_0, \tilde{J}_q) \geq \left(\frac{\bar{z}-1}{\sqrt{\bar{z}(\bar{z}-2)}}\right)^{\bar{z}} \frac{\tanh(\tilde{J}_0)+1}{4(\bar{z}-1)\tanh(\tilde{J}_0)-4}, \quad \text{for } \tilde{J}_{\text{crit}}^{\text{BG}} \leq \tilde{J}_q < \tilde{J}_0. \tag{6.5}$$

6.5.2 *Lower Bound for Quenches in the One-Phase Domain*

For quenches below the critical coupling, i.e. $\tilde{J}_q < \tilde{J}_{\rm crit}^{\rm BG}$, we have $-\tilde{\rm f}''_{\rm BG}(0, \tilde{J}_q)/\tilde{\rm f}''_{\rm BG}(0, \tilde{J}_0) > 0$, and therefore we can apply the inequality $\ln(1+x) > 2x/(2+x)$ for $x > 0$ [97] to the numerator of Eq. (6.4). Minimizing the result with respect to $\tilde{J}_q$ then yields a speed limit on the critical time, which reads

$$t_c^{\rm BG}(\tilde{J}_0, \tilde{J}_q) > \frac{\cosh^{\bar{z}}\left(\ln\left[e^{2\tilde{J}_{\rm crit}^{\rm BG}}(e^{-2\tilde{J}_0} + 2/\bar{z} + \Delta_{\bar{z}}(\tilde{J}_0))\right]/2\right)}{\bar{z} - e^{2\tilde{J}_{\rm crit}^{\rm BG}}\left[(\bar{z}-4)e^{-2\tilde{J}_0} - 4/\bar{z} - \bar{z}\Delta_{\bar{z}}(\tilde{J}_0)\right]}, \text{ for } \tilde{J}_q < \tilde{J}_{\rm crit}^{\rm BG} < \tilde{J}_0, \tag{6.6}$$

where $\Delta_{\bar{z}}(\tilde{J}_0) \equiv [8/\bar{z}^2 + e^{-4\tilde{J}_0} + (1-4/\bar{z})(1-2e^{-2\tilde{J}_0})]^{\frac{1}{2}}$. The bound becomes tighter with increasing $\tilde{J}_0$ (see inset Fig. 6.2c) and coordination number $\bar{z}$ (see Fig. 6.5b), and for $\tilde{J}_0 \to \infty$ attains a minimum value, which reads $1/8$ for $\bar{z} = 4$ (see Eq. (6.29) for the general result). Equation (6.6) is by construction smaller than Eq. (6.5), and therefore also bounds quenches within the two-phase domain. Notably, the BG critical time attains a minimum for an antiferromagnetic quench $\tilde{J}_q < \tilde{J}_{\rm Néel}^{\rm BG} < 0$ (see black point in Fig. 6.2c), which lies slightly below the Néel point $\tilde{J}_{\rm Néel}^{\rm BG} = -\tilde{J}_{\rm crit}^{\rm BG}$ [94, 95].

6.6 Antiferromagnetic Speed Limit for Relaxation

Interestingly, we now show that an antiferromagnetic speed limit also exists in the relaxation of the mean magnetization $\langle m_t \rangle \equiv \int_{-1}^{-1} m P_N(m, \tilde{J}, t) {\rm d}m$. In the thermodynamic limit $\langle m_t \rangle$ is dominated by $\bar{m}_t \equiv \arg\min_m V(m, t, \tilde{J}_q)$,[4] where $\pm\bar{m}_t$ evolve according to $\frac{\rm d}{{\rm d}t}\bar{m}_t(\tilde{J}_0, \tilde{J}_q) = 2[w^+(\bar{m}_t, \tilde{J}_q) - w^-(\bar{m}_t, \tilde{J}_q)]$, with initial condition $\bar{m}_0 = \arg\min_m \tilde{\rm f}(m, \tilde{J}_0)$ [87]. Using the **Lagrange inversion theorem**, we obtain an explicit power-series solution (see Appendix 5)

$$\bar{m}_t(\tilde{J}_0, \tilde{J}_q) = \sum_{k=0}^{\infty} \alpha_k(\tilde{J}_0, \tilde{J}_q) e^{-kt/\tau_r(\tilde{J}_q)}, \tag{6.7}$$

with relaxation rate $1/\tau_r(\tilde{J}_q) \equiv 4w^{\pm}(\bar{m}_\infty, \tilde{J}_q)\tilde{\rm f}''(\bar{m}_\infty, \tilde{J}_q)$. The prefactors α_k are given explicitly in Appendix 5. In Fig. 6.2d, e we show the relaxation of the relative excess magnetization $\delta\bar{m}_t \equiv (\bar{m}_t - m_\infty)/(\bar{m}_0 - \bar{m}_\infty)$ for quenches into the one-phase (d) and two-phase (e) domain, based on a numerical solution (lines) and by retaining only the first two terms in the power-series (dots).

---

[4] The mean magnetization $\lim_{N\to\infty}\langle m_t \rangle$ is evaluated with the saddle-point method yielding contributions from the free energy minima (see e.g. Sect.5.7.1). In the two-phase regime $\tilde{\rm f}(m, \tilde{J})$ has two global minima equidistant from $m = 0$. Therefore, $\lim_{N\to\infty}\langle m_t \rangle = 0$.

For quenches below the critical coupling, $\tilde{J}_q < \tilde{J}^{\mathrm{BG}}_{\mathrm{crit}}$, the relaxation rate depends non-monotonically on $\tilde{J}_q$ (compare red and green lines in Fig. 6.2d), which is explicitly elaborated in Fig. 6.2f. Similarly to the critical time, we find a speed limit, i.e. $\tau_r^{\mathrm{BG}}(\tilde{J}_q)$ is minimal at an antiferromagnetic quench $\tilde{J}_q^r$ below the Néel point

$$\tilde{J}_q^r \equiv \arg\min_{\tilde{J}_q} \tau_r^{\mathrm{BG}}(\tilde{J}_q) = \frac{1}{2}\ln\left(\frac{\bar{z} - 2\sqrt{\bar{z}-1}}{\bar{z}-2}\right) < \tilde{J}^{\mathrm{BG}}_{\mathrm{Néel}}\,. \tag{6.8}$$

For $\bar{z} = 4$ this gives $\tilde{J}_q^r \approx -0.65874$ as indicated in Fig. 6.2f with the black dotted line. The antiferromagnetic speed limit $\tau_r(\tilde{J}_q^r)$ is the result of a trade-off between an antiferromagnetic interaction, deterministically biasing m towards smaller values on the one hand, and growing kinetic constraints[5] on energetically accessible local configurations on the other hand.

When $\tilde{J}_q > \tilde{J}^{\mathrm{BG}}_{\mathrm{crit}}$, i.e. quenches within the two-phase regime, there is no speed limit and $\tau_r(\tilde{J}_q)$ decreases monotonically with $\tilde{J}_q$ towards zero because the distance between the initial and final location of the minima become vanishingly small, i.e. $\bar{m}_0 - \bar{m}_\infty \to 0$.

6.7 Asymptotic Measure Equivalence

Despite the presence of a cusp in the rate function for all $t > t_c$ (see [3, 4]), we now show that $P_{N\to\infty}(m, \tilde{J}, t)$ becomes **measure equivalent** [98, 99] to the equilibrium Gibbs measure exponentially fast. We quantify the distance between the two measures via the instantaneous excess free energy density $\mathscr{D}_t$ [24, 100–105] defined as the **relative entropy per spin** in the thermodynamic limit, $\mathscr{D}_t \equiv \lim_{N\to\infty} N^{-1} D[P_N(m, \tilde{J}, t)||P_N^{\mathrm{eq}}(m, \tilde{J})]$ with $D[f||g] \equiv \int f \ln(f/g)dm$, or explicitly

$$\mathscr{D}_t = \lim_{N\to\infty} \int_{-1}^{1} \mathrm{e}^{-NV(m,\tilde{J}_q,t)}[V_{\mathrm{eq}}(m, \tilde{J}_q) - V(m, \tilde{J}_q, t)]dm \simeq \sum_{k=2}^{\infty} \gamma_k(\tilde{J}_0, \tilde{J}_q)\mathrm{e}^{-kt/\tau_r(\tilde{J}_q)}, \tag{6.9}$$

where the second line was obtained with the saddle point approximation (for derivation and prefactors γ_k see Appendix 6). The time evolution of $\mathscr{D}_t$ for various quenches is shown in Fig. 6.3a. Clearly, $\mathscr{D}_{t\to\infty} \to 0$, implying that $\lim_{t\to\infty} V(m, \tilde{J}_q, t) = V_{\mathrm{eq}}(m, \tilde{J}_q)$ almost everywhere, i.e. the large deviation behavior is ergodic [98, 99].

[5] These energetic local constraints arise due to the antiferromagnetic order for $\tilde{J} < 0$.

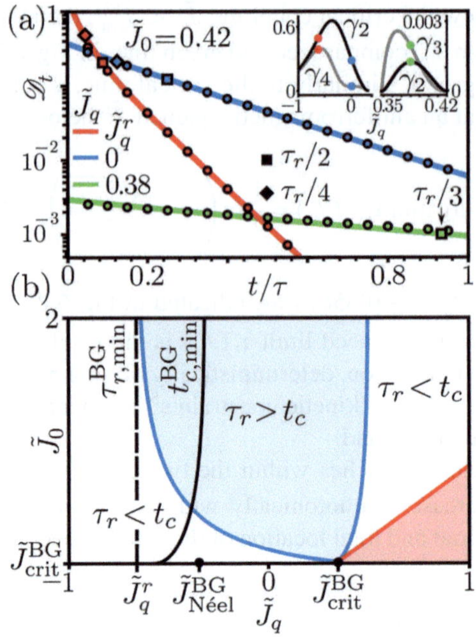

Fig. 6.3 **Relaxation of the relative entropy per spin, and the dynamical phase diagram. a** Temporal evolution of the relative entropy per spin $\mathscr{D}_t$ given by Eq. (6.9), upon a quench into the one-phase (red-blue) and two-phase regime (green). At $\tilde{J}_q^r$ (red) the relative entropy relaxes the fastest. Dots depict analytical results obtained with the first two nonzero terms in Eq. (6.9). Lines correspond to numerical results. Squares/diamonds denote the first $\tau_r/2$ and second $\tau_r/4 - \tau_r/3$ relaxation time-scales, respectively. Inset: First two nonzero prefactors which enter Eq. (6.9). **b** Dynamical phase diagram for t_c^{BG} and τ_r^{BG}. The red area is forbidden since $\tilde{J}_0 > \tilde{J}_q$. Dashed/solid black lines denote the fastest relaxation and critical time. In both panels we set $\bar{z} = 4$

6.8 Dynamical Phase Diagram

Due to the asymptotic measure equivalence, the dynamical phase transition may not always be easily observable, in particular if $t_c > \tau_r$. In Fig. 6.3b we present a **dynamical phase diagram** in the $(\tilde{J}_0, \tilde{J}_q)$-plane, showing that the critical time is not always smaller than the relaxation time. However, (i) there is an extended regime where $t_c < \tau_r$ (see blue region in Fig. 6.3b) such that the transition should be observable, and (ii) the (exact) minimal relaxation time is always smaller than the (exact) minimal critical time. The MF phase diagram is, however, starkly different (see Fig. 6.6c).

6.9 Conclusion

Our results reveal, for the first time, the finite-time dynamical phase transition in nearest-neighbor interacting Ising systems. Moreover, they unravel non-trivial **anti-ferromagnetic speed limits** for the critical time and the relaxation time of the magnetization. Considering instead quenches from antiferromagnetically ordered states, we in turn find mirror-symmetric results for the staggered magnetization (see [93–96] and Appendix 7). These unforeseen speed limits embody an optimal trade-off between antiferromagnetic interactions, biasing the magnetization towards smaller values, and a decreasing number of energetically accessible local configurations that impose kinetic constraints. As it emerges due to kinetic constraints imposed by frustrated local configurations, it should not come as a surprise that the speed limit requires accounting for nearest-neighbor correlations, and is therefore not captured by MF theory. Notably, speed limits may also be obtained from "classical" [71–77] or thermodynamic [73, 75, 78–80] speed limits which, however, is likely to be more difficult as analytical solutions for probability density functions, in particular at the critical time, do not seem to be feasible. Our findings may provide insight allowing for optimization of ultrafast optical-switching ferromagnetic materials [90]. Finally, our work provokes further intriguing questions, in particular on the microscopic path-wise understanding of the dynamical critical time, the effect of an external field, the existence of heating-cooling asymmetries [24–27] in different regimes and across phase transitions, and optimal driving protocols [20–23] that may be relevant for optical-switching ferromagnets.

Appendices

In the Appendices we present details of calculations and mathematical proofs for the claims made in this chapter. The sections are organized in the order they appear in this chapter.

Appendix 1: Hamiltonian Formalism

Recall that $V(m, \tilde{J}, t) \equiv -\lim_{N\to\infty} N^{-1} \ln(P_N(m, \tilde{J}, t))$ represents the time-dependent large-deviation rate function. In the SM of [1] it is shown that the rate function $V(m, \tilde{J}_q, t)$ with quench temperature/coupling $\tilde{J}_q$ obeys the HJ equation given by Eq. (6.2). The HJ equation can be solved with the **method of characteristics** as follows: Let $\{q(s), p(s)\}\ 0 \le s \le t$ be the characteristics that solve the **Hamilton's equations**

$$\dot{q}(s) = \partial_p \mathcal{H}(q, p), \quad \dot{p}(s) = -\partial_q \mathcal{H}(q, p), \quad q(t) = m, \quad p(0) = \tilde{\mathrm{f}}'(q(0), \tilde{J}_0), \quad (6.10)$$

where $\dot{q}(s) \equiv \mathrm{d}q(s)/\mathrm{d}s$, $\dot{p}(s) \equiv \mathrm{d}p(s)/\mathrm{d}s$, $\tilde{\mathrm{f}}'(a, \tilde{J}) \equiv \partial_m \tilde{\mathrm{f}}(m, \tilde{J})|_{m=a}$, and $\mathcal{H}(q, p)$ is given in Eq. (6.3). Upon solving the Hamilton's equations, the solution to the HJ equation reads

$$V(m, \tilde{J}_q, t) = \int_0^t [p(s)\dot{q}(s) - \mathcal{H}(q, p)]\mathrm{d}s + V(q(0), \tilde{J}_0, 0). \tag{6.11}$$

For $t > t_c$, where $t_c = t_c(\tilde{J}_0, \tilde{J}_q)$ denotes the critical time, the solutions to the Hamilton's equations become degenerate. Under these circumstances, the solution that minimizes Eq. (6.11) corresponds to the stable solution [2].

Appendix 2: Lagrangian Formalism

One can also obtain the solution to the HJ equation with the Lagrangian formalism, which is formally introduced in [3, 4]. The Lagrangian is obtained from the Hamiltonian via the **backward Legendre transform** $\mathcal{L}(q, \dot{q}) = p(q, \dot{q})\dot{q} - \mathcal{H}(q, p(q, \dot{q}))$, where $p(q, \dot{q})$ can be obtained from the first of the Hamilton's equations in Eq. (6.10) and reads

$$p(q, \dot{q}) = \frac{1}{2} \ln \left(\frac{\dot{q} + \Lambda(q, \dot{q})}{4w^+(q, \tilde{J}_q)} \right), \tag{6.12}$$

with $\Lambda(q, \dot{q}) \equiv [16w^+(q, \tilde{J}_q)w^-(q, \tilde{J}_q) + \dot{q}^2]^{1/2}$. Plugging this expression back into $\mathcal{H}(q, p(q, \dot{q}))$, we obtain the Lagrangian

$$\mathcal{L}(q, \dot{q}) = p(q, \dot{q})\dot{q} - \Lambda(q, \dot{q})/2 + w^+(q, \tilde{J}_q) + w^-(q, \tilde{J}_q). \tag{6.13}$$

The Hamilton's equations are replaced by the E-L equation, which reads

$$\begin{aligned} &\ddot{q}(s) = 2\Lambda(q, \dot{q})\partial_q[w^+(q, \tilde{J}_q) + w^-(q, \tilde{J}_q)] - 8\partial_q w^+(q, \tilde{J}_q)w^-(q, \tilde{J}_q), \\ &\dot{q}(0) = g(q(0)), \ q(t) = m. \end{aligned} \tag{6.14}$$

The boundary condition for $\dot{q}(0)$ is determined by the **curve of allowed initial configurations** (see also Eq. (24) in [4])

$$g(m) \equiv 2\mathrm{e}^{2\tilde{\mathrm{f}}'(m, \tilde{J}_0)} w^+(m, \tilde{J}_q) - 2\mathrm{e}^{-2\tilde{\mathrm{f}}'(m, \tilde{J}_0)} w^-(m, \tilde{J}_q), \tag{6.15}$$

which will be used in Sect. 6.9 to determine the **symmetry-breaking transition**. Upon solving the E-L equation, the solution of the HJ equation is given by

$$V(m, \tilde{J}_q, t) = \int_0^t \mathcal{L}(q(s), \dot{q}(s))\mathrm{d}s + V(q(0), \tilde{J}_0, 0), \tag{6.16}$$

which is identical to Eq. (6.11). Similar to the Hamiltonian formalism, the solution of Eq. (6.14) becomes degenerate for $t > t_c$. The stable solution for $q(s)$ minimizes the rate function given by Eq. (6.16).

Appendix 3: Derivation of the Critical Time

In this section we derive the critical time t_c based on two different approaches, which are discussed in [1] and [4], respectively. The first approach uses the Hamiltonian formalism to derive an equation for the curvature at $m = 0$. The second approach uses an invariance principle for the solutions of Eq. (6.14). Both approaches lead to the same result for the critical time given by Eq. (6.4). However, with the latter approach we can also derive the critical coupling above which the initial location of the cusp deviates from $m = 0$.

Appendix 3.1: Hamiltonian Formalism and the Ricatti Equation

The critical time $t_c(\tilde{J}_0, \tilde{J}_q)$ is defined as the moment when the rate function $V(m, \tilde{J}_q, t)$ develops a cusp at $m = 0$, leading to a negatively diverging curvature. In the SM of [1] an equation for the curvature $V_0''(\tilde{J}_q, t) \equiv d^2V(m, \tilde{J}_q, t)/dm^2|_{m=0}$ is derived from the Hamilton's equations. The resulting equation – after simplification – reads

$$\frac{\mathrm{d}V_0''(\tilde{J}_q, t)}{\mathrm{d}t} = 8w^{\pm}(0, \tilde{J}_q)V_0''(\tilde{J}_q, t)(\tilde{\mathrm{f}}''(0, \tilde{J}_q) - V_0''(\tilde{J}_q, t)), \tag{6.17}$$

with initial condition $V_0''(\tilde{J}_q, 0) = \tilde{\mathrm{f}}''(0, \tilde{J}_0)$. To obtain Eq. (6.17), we explicitly used the DB relation $\ln(w^-(m, \tilde{J})/w^+(m, \tilde{J})) = 2\tilde{\mathrm{f}}'(m, \tilde{J})$ and the parity symmetry $w^{\pm}(m, \tilde{J}) = w^{\mp}(-m, \tilde{J})$ to write $\partial_m w^{\pm}(m, \tilde{J})|_{m=0} = \mp w^{\pm}(0, \tilde{J})\tilde{\mathrm{f}}''(0, \tilde{J})$. Equation (6.17) is a so-called **Ricatti equation**, which can be solved analytically. The resulting solution *valid up to the critical time* reads

$$V_0''(\tilde{J}_q, t) = \frac{\tilde{\mathrm{f}}''(0, \tilde{J}_q)}{1 - (1 - \tilde{\mathrm{f}}''(0, \tilde{J}_q)/\tilde{\mathrm{f}}''(0, \tilde{J}_0))\mathrm{e}^{-2t/\hat{\tau}_r(\tilde{J}_q)}}, \tag{6.18}$$

where $1/\hat{\tau}_r(\tilde{J}_q) \equiv 4w^{\pm}(0, \tilde{J}_q)\tilde{\mathrm{f}}''(0, \tilde{J}_q)$ is an effective relaxation rate. The critical time t_c determines the root of the denominator in Eq. (6.18), which leads to Eq. (6.4).

Appendix 3.2: Lagrangian Formalism and the Cusp Location

Following the steps in Sect. 3.5 of [4] we can derive the **symmetry-breaking coupling** $\tilde{J}_0^{\mathrm{SB}}(\tilde{J}_q)$, above which the initial location of the cusp deviates from $\bar{m} = 0$. The idea behind this calculation is that at the critical time the solution of Eq. (6.14) converges to the same point $q(t_c)$ for different initial conditions $\{q(0), \dot{q}(0)\}$. In other words, *the location of $q(t_c)$ remains invariant under a variation of the initial conditions*. To determine the symmetry-breaking transition, it suffices to consider the dynamics of $q(s)$ around the origin [4]. We linearize Eq. (6.14) around the point $(q, \dot{q}) = (0, 0)$, which yields

$$\ddot{q}(s) = q(s)/\hat{\tau}_r^2(\tilde{J}_q), \ \dot{q}(0) = g(q_0) \equiv v_0, \ q(0) \equiv q_0, \tag{6.19}$$

where $\{q_0, v_0(q_0)\}$ are the initial conditions, and $1/\hat{\tau}_r(\tilde{J}_q) \equiv 4w^{\pm}(0, \tilde{J}_q)\tilde{\mathrm{f}}''(0, \tilde{J}_q)$. The solution of Eq. (6.19) is given by

$$q(s) = (q_0/2 - \hat{\tau}_r v_0/2)\mathrm{e}^{-s/\hat{\tau}_r} + (q_0/2 + \hat{\tau}_r v_0/2)\mathrm{e}^{s/\hat{\tau}_r}. \tag{6.20}$$

We now consider a variation of $q(s)$ w.r.t. the initial conditions $\{q_0, v_0(q_0)\}$, which gives

$$\begin{aligned} \frac{\mathrm{d}q(s)}{\mathrm{d}q_0} &= \frac{\partial q(s)}{\partial q_0} + \frac{\partial q(s)}{\partial v_0} g'(q_0) \\ &= (1/2 - \hat{\tau}_r g'(q_0)/2)\mathrm{e}^{-s/2\hat{\tau}_r} + (1/2 + \hat{\tau}_r g'(q_0)/2)\mathrm{e}^{s/2\hat{\tau}_r}, \end{aligned} \tag{6.21}$$

where $g'(q_0) \equiv \mathrm{d}g(m)/\mathrm{d}m|_{m=q_0}$ and $g(m)$ is the *curve of allowed initial configurations* given by Eq. (6.15). At the critical time $s = t_c$ the variation in Eq. (6.21) should vanish, which leads to the critical time in the form

$$t_c(q_0) = (\hat{\tau}_r/2) \ln\left(\frac{g'(q_0) - 1/\hat{\tau}_r}{g'(q_0) + 1/\hat{\tau}_r}\right). \tag{6.22}$$

For $\tilde{J}_c < \tilde{J}_0 < \tilde{J}_0^{\mathrm{SB}}(\tilde{J}_q)$ the critical time given by Eq. (6.22) has a single minimum at $q_{\min} = 0$ (see upper line in Fig. 6.4a). Inserting $q_0 = 0$ and recalling the relation $\partial_m w^{\pm}(m, \tilde{J})|_{m=0} = \mp w^{\pm}(0, \tilde{J})\tilde{\mathrm{f}}''(0, \tilde{J})$ we obtain the critical time given by Eq. (6.4).

For $\tilde{J}_0 > \tilde{J}_0^{\mathrm{SB}}(\tilde{J}_q)$ Eq. (6.22) develops two minima at $\pm q_{\min} \neq 0$, corresponding to the new cusp locations (see lower line in Fig. 6.4a).

For $\tilde{J}_0 = \tilde{J}_0^{\mathrm{SB}}(\tilde{J}_q)$ the curvature of Eq. (6.22) at $q_0 = 0$ vanishes (see middle line in Fig. 6.4a), which results in the following equation determining $\tilde{J}_0^{\mathrm{SB}}(\tilde{J}_q)$

$$g'''(0)|_{\tilde{J}_0^{\mathrm{SB}}(\tilde{J}_q)} = 0, \tag{6.23}$$

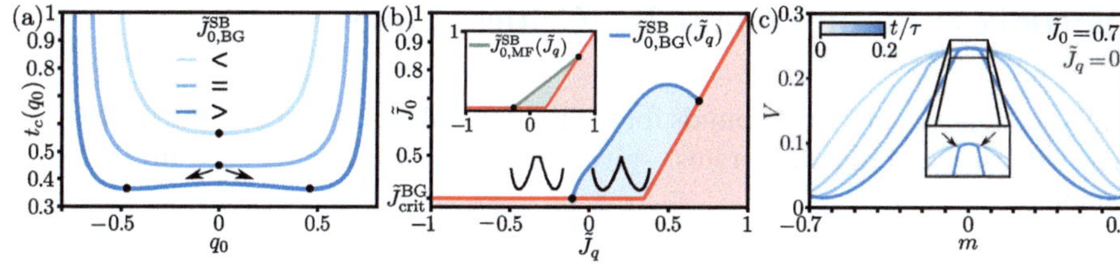

Fig. 6.4 Symmetry-breaking transition for the location of the cusp. In all panels we consider a lattice with $\bar{z} = 4$. **a** BG critical time $t_c^{\mathrm{BG}}(q_0)$ given by Eq. (6.22) as a function of the initial point q_0 for various values of $\tilde{J}_0$. The black dots indicate the minima of $t_c^{\mathrm{BG}}(q_0)$, which set the location of the cusp. For $\tilde{J}_0 > \tilde{J}_{0,\mathrm{BG}}^{\mathrm{SB}}$ the critical time contains two minima $\pm q_{\min}$, which correspond to non-zero cusp locations. **b** Blue line: BG symmetry-breaking coupling $\tilde{J}_{0,\mathrm{BG}}^{\mathrm{SB}}(\tilde{J}_q)$ given by Eq. (6.25) as a function of $\tilde{J}_q$. Inside the light blue region the cusp is formed at $m = 0$, and in the white region the cusp is formed at $m \neq 0$. The red area is forbidden since $\tilde{J}_0 > \tilde{J}_{\mathrm{crit}}^{\mathrm{BG}}$ and $\tilde{J}_0 > \tilde{J}_q$. Inset: MF symmetry-breaking coupling $\tilde{J}_0^{\mathrm{MF}}(\tilde{J}_q)$ given by Eq. (6.24). Inside the light green region the cusp is formed at $m = 0$. **c** Temporal evolution of the BG rate function $V_{\mathrm{BG}}(m, \tilde{J}_q, t)$ for a quench to $\tilde{J}_q = 0$. Time increases from light to dark blue. The initial coupling is set above the symmetry-breaking coupling, i.e. $\tilde{J}_0 > \tilde{J}_0^{\mathrm{SB}}(\tilde{J}_q)$, to induce a cusp at $m \neq 0$. Inset: Enlargement of the rate function around the center. Black arrows indicate the location of the cusps

where we have used $g''(0) = 0$. Solving Eq. (6.23) for the MF approximation, we obtain

$$\tilde{J}_{0,\mathrm{MF}}^{\mathrm{SB}}(\tilde{J}_q) = \frac{3 + \bar{z}\tilde{J}_q}{2\bar{z}}. \tag{6.24}$$

For $\tilde{J}_q = 0$ we have $\tilde{J}_{0,\mathrm{MF}}^{\mathrm{SB}}(0) = 3/2z$ as mentioned in [3, 4]. Within the BG approximation the general formula for $\tilde{J}_{0,\mathrm{BG}}^{\mathrm{SB}}(\tilde{J}_q)$ is rather long and therefore not shown. For $\bar{z} = 4$ the result can compactly be written as

$$\tilde{J}_{0,\mathrm{BG}}^{\mathrm{SB}}(\tilde{J}_q)|_{\bar{z}=4} = \ln(x_{\tilde{J}_q})/2, \tag{6.25}$$

where $x_{\tilde{J}_q}$ is the *real solution* of the following cubic equation

$$\begin{aligned} &20 - 16(1 + 2\mathrm{e}^{-2\tilde{J}_q})x_{\tilde{J}_q} + (8 + 8\mathrm{e}^{-2\tilde{J}_q} + 20\mathrm{e}^{-4\tilde{J}_q})x_{\tilde{J}_q}^2 \\ &\quad - (2 - 4\mathrm{e}^{-2\tilde{J}_q} + 10\mathrm{e}^{-4\tilde{J}_q} - \mathrm{e}^{-8\tilde{J}_q} + 6\mathrm{e}^{-10\tilde{J}_q} - 9\mathrm{e}^{-12\tilde{J}_q} + 4\mathrm{e}^{-14\tilde{J}_q})x_{\tilde{J}_q}^3 = 0. \end{aligned} \tag{6.26}$$

For $\tilde{J}_q = 0$ we obtain $\tilde{J}_{0,\mathrm{BG}}^{\mathrm{SB}}(0) = \ln([\bar{z} + 1][\bar{z} - 2])/2$ as mentioned in Sect. 6.3. In Fig. 6.4b we plot Eq. (6.25) as a function of $\tilde{J}_0$ with the dark blue line. Interestingly, the light blue region for which the cusp appears at $m = 0$ is rather small and of finite area. Correspondingly, in Fig. 6.4c we provide an example of the rate function $V_{\mathrm{BG}}(m, \tilde{J}_q, t)$ for which the cusps appear at a non-zero location during the evolution towards to equilibrium state.

Appendix 4: Bounds on the BG Critical Time

In this section we derive the bounds for the BG critical time t_c^{BG}. Inserting the BG free energy density and transition rates—given by Eqs. (2.14) and (6.1)—into Eq. (6.4), we obtain

$$t_c^{\mathrm{BG}}(\tilde{J}_0, \tilde{J}_q) = \frac{\cosh^{\bar{z}}(\tilde{J}_q)(\tanh(\tilde{J}_q)+1)}{4(\bar{z}-1)\tanh(\tilde{J}_q)-4}\left[\tilde{J}_q + \ln\left(\frac{(\bar{z}-1)\sinh(\tilde{J}_0)-\cosh(\tilde{J}_0)}{\bar{z}\sinh(\tilde{J}_0-\tilde{J}_q)}\right)\right], \quad (6.27)$$

where $\tilde{J}_0 > \tilde{J}_{\mathrm{crit}}^{\mathrm{BG}} \equiv \ln(\bar{z}/(\bar{z}-2))/2$ and $\tilde{J}_0 > \tilde{J}_q$. Figure 6.2c displays the BG critical given by Eq. (6.27) with the blue line. The BG critical time has a minimum for an antiferromagnetic quench $\tilde{J}_q < 0$, which cannot be determined analytically. We can, however, derive lower bounds on the critical time. To construct the bounds, we will distinguish between quenches in the one- and two-phase domain, i.e. $\tilde{J}_q < \tilde{J}_{\mathrm{crit}}^{\mathrm{BG}}$ and $\tilde{J}_q \geq \tilde{J}_{\mathrm{crit}}^{\mathrm{BG}}$.

Appendix 4.1: $\tilde{J}_q < \tilde{J}_{\mathrm{crit}}^{\mathrm{BG}}$

For quenches in the one-phase domain we can bound the critical time by applying the well-known inequality $\ln(1+x) > 2x/(2+x)$ for $x > 0$ [97] to the logarithmic term in Eq. (6.4) (since $-\tilde{\mathrm{f}}_{\mathrm{BG}}''(0, \tilde{J}_q)/\tilde{\mathrm{f}}_{\mathrm{BG}}''(0, \tilde{J}_0) > 0$). This yields the *local* lower bound

$$t_c^{\dagger\mathrm{BG}}(\tilde{J}_0, \tilde{J}_q) = \frac{\cosh^{\bar{z}}(\tilde{J}_q)}{\bar{z}-2+\bar{z}\mathrm{e}^{-2\tilde{J}_q}-2\bar{z}\mathrm{e}^{-2\tilde{J}_0}}. \quad (6.28)$$

In Fig. 6.5a we plot $t_c^{\dagger\mathrm{BG}}$ with the black line. Surprisingly, this local bound also seems to work for $\tilde{J}_q \geq \tilde{J}_{\mathrm{crit}}^{\mathrm{BG}}$, even though $-\tilde{\mathrm{f}}_{\mathrm{BG}}''(0, \tilde{J}_q)/\tilde{\mathrm{f}}_{\mathrm{BG}}''(0, \tilde{J}_0) < 0$. Furthermore, it gives the exact result for $\tilde{J}_q = \tilde{J}_{\mathrm{crit}}^{\mathrm{BG}}$ given by Eq. (6.5). The lower bound is also non-monotonic w.r.t. $\tilde{J}_q$, and displays a minimum for an anti-ferromagnetic quench $\tilde{J}_q < 0$. At the respective minimum, the global lower bound $\inf_{\tilde{J}_q} t_c^{\dagger\mathrm{BG}}(\tilde{J}_0, \tilde{J}_q)$ (see black dashed line in Fig. 6.5a) is given by Eq. (6.6). Taking the limit $\tilde{J}_0 \to \infty$ of Eq. (6.6), we further obtain the following universal global lower bound independent of $\tilde{J}_q$ and $\tilde{J}_0$ that reads

$$\lim_{\tilde{J}_0\to\infty} \inf_{\tilde{J}_q} t_c^{\dagger\mathrm{BG}}(\tilde{J}_0, \tilde{J}_q) = \frac{(\bar{z}-2)^{1-\bar{z}/2}[2+\nu_{\bar{z}}]^{-\bar{z}/2}[\bar{z}+\nu_{\bar{z}}]^{\bar{z}}}{2^{\bar{z}}(4+\bar{z}[\bar{z}-2+\nu_{\bar{z}}])}, \quad (6.29)$$

with $\nu_{\bar{z}} \equiv \sqrt{8+\bar{z}(\bar{z}-4)}$. For $\bar{z}=4$ this gives the universal global lower bound $t_c^{\mathrm{BG}}(\tilde{J}_0, \tilde{J}_q) > 1/8$ and is shown with the red line in Fig. 6.5a. In Fig. 6.5b we observe that for increasing $\bar{z}$ the bounds given by Eq. (6.6) and Eq. (6.29) become sharper with respect to the true/exact minimum of t_c^{BG}.

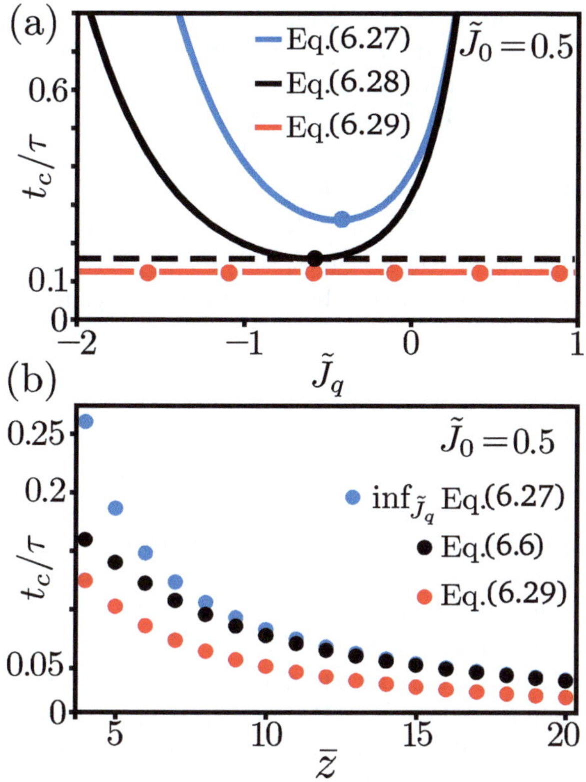

Fig. 6.5 Bounds on the BG critical time for quenches in the one-phase domain. a BG critical time $t_c^{\mathrm{BG}}(\tilde{J}_0, \tilde{J}_q)$ given by Eq. (6.27) (blue line) as a function of $\tilde{J}_q$ for $\tilde{J}_0 = 0.5$ and $\bar{z} = 4$. The respective lower bounds are shown with the black and red line. **b** Minimum of the BG critical time (blue dots) as a function of the lattice coordination number $\bar{z}$. The respective lower bounds are shown with the black and red dots, respectively

Appendix 4.2: $\tilde{J}_q \geq \tilde{J}_{\mathrm{crit}}^{\mathrm{BG}}$

For quenches in the two-phase domain we prove that the BG critical time $t_c^{\mathrm{BG}}(\tilde{J}_0, \tilde{J}_q)$ is bounded from below by the critical quench $t_c^{\mathrm{BG}}(\tilde{J}_0, \tilde{J}_{\mathrm{crit}}^{\mathrm{BG}})$, which is given by Eq. (6.5). To prove that Eq. (6.5) provides a lower bound for the critical time for quenches in the two-phase domain, we first differentiate Eq. (6.27) w.r.t. $\tilde{J}_q$, which gives

$$\frac{\partial t_c^{\mathrm{BG}}(\tilde{J}_0, \tilde{J}_q)}{\partial \tilde{J}_q} = \frac{\bar{z}(1 + \tanh(\tilde{J}_q)) \cosh^{\bar{z}}(\tilde{J}_q)}{4((\bar{z} - 1)\tanh(\tilde{J}_q) - 1)^2} \mathcal{A}_1(\tilde{J}_0, \tilde{J}_q), \tag{6.30}$$

where we have introduced the auxiliary function (and subsequent auxiliary functions)

$$\begin{aligned} \mathcal{A}_1(\tilde{J}_0, \tilde{J}_q) &\equiv -\mathcal{A}_2(\tilde{J}_0, \tilde{J}_q)[1 - \tanh(\tilde{J}_q)] - \mathcal{A}_3(\tilde{J}_q) \ln(\mathcal{A}_2(\tilde{J}_q, \tilde{J}_0)), \\ \mathcal{A}_2(\tilde{J}_0, \tilde{J}_q) &\equiv [(\bar{z}-1)\tanh(\tilde{J}_q) - 1][1 + \tanh(\tilde{J}_0)]/[\bar{z}(\tanh(\tilde{J}_q) - \tanh(\tilde{J}_0))], \\ \mathcal{A}_3(\tilde{J}_q) &\equiv 1 - (\bar{z}-1)\tanh^2(\tilde{J}_q). \end{aligned} \tag{6.31}$$

All terms in front of $\mathcal{A}_1(\tilde{J}_0, \tilde{J}_q)$ in Eq. (6.30) are trivially positive. If furthermore $\mathcal{A}_1(\tilde{J}_0, \tilde{J}_q) > 0$ for $\tilde{J}^{\text{BG}}_{\text{crit}} < \tilde{J}_q < \tilde{J}_0$, then we know that Eq. (6.5) provides a lower bound. To prove that the latter is positive, we proceed in two steps.

Step 1: $\mathcal{A}_2(\tilde{J}_q, \tilde{J}_0) > 1,\ \forall \tilde{J}_q > \tilde{J}^{\text{BG}}_{\text{crit}}$

First, we focus on the term $\mathcal{A}_2(\tilde{J}_q, \tilde{J}_0)$ entering the logarithm in $\mathcal{A}_1(\tilde{J}_0, \tilde{J}_q)$. Here we prove that $\mathcal{A}_2(\tilde{J}_q, \tilde{J}_0) > 1,\ \forall \tilde{J}_q > \tilde{J}^{\text{BG}}_{\text{crit}}$, which we need for the second step. Note that $\mathcal{A}_2(\tilde{J}^{\text{BG}}_{\text{crit}}, \tilde{J}_0) = 1$, which can easily be checked by hand. Introducing $x_0 \equiv \tanh(\tilde{J}_0)$ and $x_q \equiv \tanh(\tilde{J}_q)$, we find $\partial_{\tilde{J}_q} \mathcal{A}_2(\tilde{J}_q, \tilde{J}_0) = \cosh^{-2}(\tilde{J}_q) \partial_{x_q} \mathcal{A}_2(x_q, x_0) > 0,\ \forall \tilde{J}_0 > \tilde{J}^{\text{BG}}_{\text{crit}}$. To see this, we write out the partial derivative and obtain

$$\begin{aligned} \partial_{x_q} \mathcal{A}_2(x_q, x_0) &= \partial_{x_q} \left(\frac{(1+x_q)[(\bar{z}-1)x_0-1]}{\bar{z}(x_0-x_q)} \right) \\ &= \frac{(1+x_0)((\bar{z}-1)x_0-1)}{\bar{z}(x_0-x_q)^2} > 0,\ \forall x_0 > (\bar{z}-1)^{-1}. \end{aligned} \tag{6.32}$$

Finally, note that $x_0 > (\bar{z}-1)^{-1}$ translates to $\tilde{J}_0 > \operatorname{arctanh}(\bar{z}-1)^{-1} = \tilde{J}^{\text{BG}}_{\text{crit}}$, which is the regime of interest. Hence, $\mathcal{A}_2(\tilde{J}_q, \tilde{J}_0)$ has a positive slope w.r.t. $\tilde{J}_q$. Combined with $\mathcal{A}_2(\tilde{J}^{\text{BG}}_{\text{crit}}, \tilde{J}_0) = 1$, this proves that $\mathcal{A}_2(\tilde{J}_q, \tilde{J}_0) > 1,\ \forall \tilde{J}_q > \tilde{J}^{\text{BG}}_{\text{crit}}$.

Step 2: $\mathcal{A}_1(\tilde{J}_0, \tilde{J}_q) > 0,\ \forall \tilde{J}_q > \tilde{J}^{\text{BG}}_{\text{crit}}$

Now we turn our attention to $\mathcal{A}_1(\tilde{J}_0, \tilde{J}_q)$. We begin by considering the regime $\tanh(\tilde{J}_q) \geq 1/\sqrt{z-1}$. Here $\mathcal{A}_3(\tilde{J}_q) < 0$, and therefore $-\mathcal{A}_3(\tilde{J}_q) \ln(\mathcal{A}_2(\tilde{J}_q, \tilde{J}_0)) > 0$ based on the previous step. Furthermore, $-\mathcal{A}_2(\tilde{J}_0, \tilde{J}_q)[1 - \tanh(\tilde{J}_q)] > 0,\ \forall \tilde{J}^{\text{BG}}_{\text{crit}} < \tilde{J}_q < \tilde{J}_0$, and so it follows that $\mathcal{A}_1(\tilde{J}_0, \tilde{J}_q) > 0$ for $1/\sqrt{z-1} \leq \tanh(\tilde{J}_q) \leq \tanh(\tilde{J}_0)$.

Next we consider the regime $1/(z-1) < \tanh(\tilde{J}_q) < 1/\sqrt{z-1}$. Here $\mathcal{A}_3(\tilde{J}_q) > 0$, and therefore $-\mathcal{A}_3(\tilde{J}_q) \ln(\mathcal{A}_2(\tilde{J}_q, \tilde{J}_0)) < 0$. To construct a bound for $\mathcal{A}_1(\tilde{J}_0, \tilde{J}_q)$, we apply the following chain of inequalities

$$\begin{aligned} \mathcal{A}_1(\tilde{J}_0, \tilde{J}_q) &\equiv \mathcal{A}_3(\tilde{J}_q)[-\mathcal{A}_2(\tilde{J}_0, \tilde{J}_q)[1 - \tanh(\tilde{J}_q)]/\mathcal{A}_3(\tilde{J}_q) - \ln(\mathcal{A}_2(\tilde{J}_q, \tilde{J}_0))] \\ &> \mathcal{A}_3(\tilde{J}_q)[-\mathcal{A}_2(\tilde{J}_0, \tilde{J}_q)[1 - \tanh(\tilde{J}_q)]/\mathcal{A}_3(\tilde{J}_q) - \mathcal{A}_2(\tilde{J}_q, \tilde{J}_0) + 1] \\ &> \mathcal{A}_3(\tilde{J}_q)[-\mathcal{A}_2(\tilde{J}_0, \tilde{J}_q) - \mathcal{A}_2(\tilde{J}_q, \tilde{J}_0) + 1] = 0. \end{aligned}$$

In passing from the first to the second line, we have applied the inequality $\ln(z) < z - 1$ for $z > 1$. From the second to the third line we have used $[1 - \tanh(\tilde{J}_q)]/\mathcal{A}_3(\tilde{J}_q) > 1$ for $1/(z-1) < \tanh(\tilde{J}_q) < 1/\sqrt{z-1}$. Finally, in the last line we used that $1 - \mathcal{A}_2(x_0, x_q) - \mathcal{A}_2(x_q, x_0) = 0$, which follows by simply writing out the terms.

Combining the results, we find that $\mathcal{A}_1(\tilde{J}_0, \tilde{J}_q) > 0$ for $\tilde{J}^{\text{BG}}_{\text{crit}} < \tilde{J}_q < \tilde{J}_0$, and therefore $t_c^{\text{BG}}(\tilde{J}_0, \tilde{J}_q)$ is bounded by Eq. (6.5) in this regime.

Appendix 5: Relaxation Dynamics

In this section we focus on the relaxation dynamics of the rate function minima, $\bar{m}(t, \tilde{J}_0, \tilde{J}_q) \equiv \arg\min_m V(m, \tilde{J}_q, t)$. Based on the first characteristic equation in Eq. (6.10) we find that the minima obey the differential equation

$$\frac{\mathrm{d}\bar{m}(t, \tilde{J}_0, \tilde{J}_q)}{\mathrm{d}t} = 2w^+(\bar{m}, \tilde{J}_q) - 2w^-(\bar{m}, \tilde{J}_q). \tag{6.33}$$

As the RHS does not depend explicitly on time, the solution is given by the integral

$$\frac{1}{2}\int \frac{\mathrm{d}\bar{m}}{w^+(\bar{m}, \tilde{J}_q) - w^-(\bar{m}, \tilde{J}_q)} = t + \mathcal{C}, \tag{6.34}$$

where $\mathcal{C} = \mathcal{C}(\tilde{J}_0, \tilde{J}_q)$ is an integration constant left to be determined from the initial condition at $t = 0$. The integral on the LHS cannot be evaluated analytically upon inserting the MF transition rates (see Eq. (3) in [1] for their functional form). However, for the BG transition rates given by Eq. (6.1), the integral can be evaluated explicitly for $\bar{z} = \{2, 3, 4, 5, 6\}$. Here we show the analysis for $\bar{z} = \{2, 4\}$, where we use the former as an educative introduction to carry out the latter. Our aim is to go beyond the linear response regime studied in [87] by applying the so-called Lagrange Inversion Theorem.

Appendix 5.1: BG Approximation with $\bar{z} = 2$

Formally, the mean magnetization for $\bar{z} = 2$ vanishes for any initial and final temperature/coupling. However, instead of considering a temperature/coupling quench, we consider a magnetization quench where we initially prepare the system in a non-zero magnetic state with $\bar{m}(0) \equiv \bar{m}_0 \neq 0$. Inserting the BG transition rates with $\bar{z} = 2$ into Eq. (6.34), we obtain – after some algebraic manipulation – the result

$$-\tau_r(\tilde{J}_q)\ln(\bar{m}/f(\bar{m}, \tilde{J}_q)) = t + \mathcal{C}, \tag{6.35}$$

where $1/\tau_r(\tilde{J}_q) \equiv 4w^{\pm}_{\mathrm{BG}}(0, \tilde{J}_q)\tilde{\mathrm{f}}''_{\mathrm{BG}}(0, \tilde{J}_q) = 8/(1 + \mathrm{e}^{2\tilde{J}_q})^2$ is the relaxation rate for $\bar{z} = 2$, and we have introduced the auxiliary function

$$f(\bar{m}, \tilde{J}_q) \equiv \exp(-\tanh(\tilde{J}_q)\ln(\alpha_+) - \alpha_-/[2\cosh(\tilde{J}_q)\bar{m}]^2), \tag{6.36}$$

with

$$\alpha_\pm(\bar{m}, \tilde{J}_q) \equiv \exp(2\tilde{J}_q) \pm [\bar{m}^2 + \exp(4\tilde{J}_q)(1-\bar{m}^2)]^{1/2}. \tag{6.37}$$

From Eq. (6.35) we directly read off the integration constant $\mathcal{C} = \mathcal{C}(\bar{m}_0, \tilde{J}_q)$ at $t = 0$. To obtain an explicit solution for $\bar{m}$, we multiply both sides of Eq. (6.35) by $-\tau_r$, and subsequently exponentiate, resulting in

$$\frac{\bar{m}}{f(\bar{m}, \tilde{J}_q)} = \frac{\bar{m}_0}{f(\bar{m}_0, \tilde{J}_q)} \mathrm{e}^{-t/\tau_r(\tilde{J}_q)}, \tag{6.38}$$

where we have now also fixed the integration constant. Now we invoke the *Lagrange inversion theorem*: Let $h(w)$ be a function which is analytic in some neighborhood of the point $w = 0$ (of the complex plane) with $h(0) \neq 0$, and let it satisfy the equation

$$\frac{w}{h(w)} = \xi. \tag{6.39}$$

Then $\exists a, b \in \mathbb{R}^+$ such that for $|\xi| < a$ Eq. (6.39) has only a single solution in the domain $|w| < b$. According to the **Lagrange-Bürmann** formula this unique solution is an analytical function of ξ given by

$$w = \sum_{k=1}^{\infty} \frac{\xi^k}{k!} \left[\frac{d^{k-1}}{dw^{k-1}} h(w)^k \right]_{w=0}. \tag{6.40}$$

Note that Eq. (6.38) is similar in structure to Eq. (6.39), and furthermore

$$f(0, \tilde{J}_q) = \exp\left(-\tanh(\tilde{J}_q)(1/2 + \ln 2 + 2\tilde{J}_q)\right), \tag{6.41}$$

which is non-zero $\forall \tilde{J}_q \in \mathbb{R}$. Therefore, we can use Eq. (6.40) to obtain an explicit solution for $\bar{m}$, yielding

$$\begin{aligned} \bar{m}(t, \bar{m}_0, \tilde{J}_q) &= \sum_{k=1}^{\infty} \frac{\bar{m}_0^k}{f(\bar{m}_0, \tilde{J}_q)^k k!} \left[\frac{d^{k-1} f(\bar{m}, \tilde{J}_q)^k}{dw^{k-1}} \right]_{\bar{m}=0} \mathrm{e}^{-kt/\tau_r(\tilde{J}_q)} \\ &= \sum_{k=1}^{\infty} \alpha_k(\bar{m}_0, \tilde{J}_q) \mathrm{e}^{-kt/\tau_r(\tilde{J}_q)}. \end{aligned} \tag{6.42}$$

For completeness, we list the first three non-zero coefficients

$$\begin{aligned} \alpha_1(\bar{m}_0, \tilde{J}_q) &= \bar{m}_0 f(0, \tilde{J}_q)/f(\bar{m}_0, \tilde{J}_q), \\ \alpha_3(\bar{m}_0, \tilde{J}_q) &= \alpha_1^3(\bar{m}_0, \tilde{J}_q) \mathrm{e}^{-4\tilde{J}_q}(1 - \mathrm{e}^{2\tilde{J}_q})^2/8, \\ \alpha_5(\bar{m}_0, \tilde{J}_q) &= \alpha_1^5(\bar{m}_0, \tilde{J}_q) \mathrm{e}^{-4\tilde{J}_q} \sinh(\tilde{J}_q)^3 (4\cosh(\tilde{J}_q) + 5\sinh(\tilde{J}_q))/8. \end{aligned} \tag{6.43}$$

Note that $\alpha_1(\bar{m}_0, 0) = \bar{m}_0$ and $\alpha_k(\bar{m}_0, 0) = 0, \ \forall k \in \{2, 3, \ldots\}$, which gives the well-known result $\bar{m}(t, \bar{m}_0, 0) = \bar{m}_0 \exp(-2t)$ [4]. Furthermore, since $f(\bar{m}, \tilde{J}_q) = f(-\bar{m}, \tilde{J}_q)$, we know that $\alpha_{2k} = 0, \ \forall k \in \mathbb{N}$. This concludes our derivation of $\bar{m}(t, \bar{m}_0, \tilde{J}_q)$ for $\bar{z} = 2$.

Appendix 5.2: BG Approximation with $\bar{z} = 4$

Now we focus on the case $\bar{z} = 4$. The analysis requires the same steps as shown in the previous section, but involves a bit more algebra. We will focus only on quenches where the initial coupling is above the critical coupling, i.e. $\tilde{J}_0 > \tilde{J}_{\text{crit}}^{\text{BG}} = \ln(2)/2$, resulting in the following initial magnetization (see Eq. (2.37) and recall that $\varphi = (1-m)/2$)

$$\bar{m}_0(\tilde{J}_0) = \mathrm{e}^{2\tilde{J}_0}(\mathrm{e}^{4\tilde{J}_0} - 4)^{1/2}/(\mathrm{e}^{4\tilde{J}_0} - 2). \tag{6.44}$$

In order to apply the *Lagrange inversion theorem*, we have to make a distinction between quenches above and below the critical coupling, since they have different equilibrium states. Furthermore, for quenches below the critical coupling $\tilde{J}_q \leq \ln(2)/2$, we will encounter a particular "special" value $\tilde{J}_q = \ln(2)/4$ which needs to be handled separately.

Regime 1. $\tilde{J}_q < \ln(2)/2$ and $\tilde{J}_q \neq \ln(2)/4$

Upon determining the integral in Eq. (6.34) for $\bar{z} = 4$, we obtain an analytic expression which can be written in a similar form as Eq. (6.35). In the one-phase domain the relaxation rate is given by $1/\tau_r(\tilde{J}_q) \equiv 4w_{\text{BG}}^{\pm}(0, \tilde{J}_q)\tilde{\mathrm{f}}_{\text{BG}}''(0, \tilde{J}_q) = \cosh^4(\tilde{J}_q)/(4\exp(-2\tilde{J}_q) - 2)$, which is plotted in Fig. 6.2f (blue line, left side of the critical coupling). The auxiliary function $f(\bar{m}, \tilde{J}_q)$ in Eq. (6.35) is now given by

$$f(\bar{m}, \tilde{J}_q) = \prod_{i=1}^{5} f_i(\bar{m}, \tilde{J}_q), \tag{6.45}$$

which we have further divided into sub-auxiliary functions that read

$$\begin{aligned}
f_1(\bar{m}, \tilde{J}_q) &= \exp\left(\frac{\alpha_- \operatorname{sech}(\tilde{J}_q)^6(1 - 3\tanh(\tilde{J}_q))(2 + \mathrm{e}^{2\tilde{J}_q})^2}{8\bar{m}^4(\tanh(\tilde{J}_q) - 3)^3}\right), \\
f_2(\bar{m}, \tilde{J}_q) &= \exp\left(\frac{\mathrm{e}^{2\tilde{J}_q}(2 - \mathrm{e}^{2\tilde{J}_q})(2\alpha_- + (13\alpha_- - 2)\mathrm{e}^{2\tilde{J}_q} + (5\alpha_- + 1)\mathrm{e}^{4\tilde{J}_q} + \mathrm{e}^{6\tilde{J}_q})}{(1 + \mathrm{e}^{2\tilde{J}_q})^3(2 + \mathrm{e}^{2\tilde{J}_q})^2\bar{m}^2}\right), \\
f_3(\bar{m}, \tilde{J}_q) &= \alpha_+(\bar{m}, \tilde{J}_q)^{\nu_1(\tilde{J}_q)}, \\
f_4(\bar{m}, \tilde{J}_q) &= [4\bar{m}^2 - \mathrm{e}^{4\tilde{J}_q}(\mathrm{e}^{4\tilde{J}_q} - 4)(1 - \bar{m}^2)]^{\nu_2(\tilde{J}_q)}, \\
f_5(\bar{m}, \tilde{J}_q) &= [4\bar{m}^2 + \mathrm{e}^{4\tilde{J}_q}((2 - \alpha_+ + \mathrm{e}^{2\tilde{J}_q} - \mathrm{e}^{4\tilde{J}_q})^2 - \bar{m}^2(3 - \mathrm{e}^{4\tilde{J}_q})^2)]^{-\nu_2(\tilde{J}_q)/2},
\end{aligned} \tag{6.46}$$

and $\alpha_\pm(\bar{m}, \tilde{J}_q)$ is given by Eq. (6.37). The exponents in the last three equations read by

$$\begin{aligned}
\nu_1(\tilde{J}_q) &\equiv [\operatorname{sech}(\tilde{J}_q)^4(3x_q - 1) + \operatorname{sech}(\tilde{J}_q)^2(19x_q - 11) + 44x_q - 20](x_q - 3)^{-3}, \\
\nu_2(\tilde{J}_q) &\equiv 32\mathrm{e}^{2\tilde{J}_q}(2 + \mathrm{e}^{2\tilde{J}_q})^{-3}(\mathrm{e}^{4\tilde{J}_q} - 2)^{-1},
\end{aligned} \tag{6.47}$$

with $x_q \equiv \tanh(\tilde{J}_q)$. Note that $\nu_2 \to \infty$ for $\tilde{J}_q \to \ln(2)/4$. This coupling value is a particular point where the integral Eq. (6.34) drastically simplifies as we will see in the next section. To check whether we can apply the *Lagrange inversion theorem*, we first need to determine $f(0, \tilde{J}_q)$, which reads

$$f(0, \tilde{J}_q) = 2^{\nu_1(\tilde{J}_q)-\nu_2(\tilde{J}_q)} e^{2[\nu_1(\tilde{J}_q)+\nu_2(\tilde{J}_q)]\tilde{J}_q - \nu_3(\tilde{J}_q)} |\coth(\tilde{J}_q) - 3|^{\nu_2(\tilde{J}_q)}, \tag{6.48}$$

where we have defined the auxiliary function

$$\nu_3(\tilde{J}_q) = (9e^{8\tilde{J}_q} - 2e^{6\tilde{J}_q} - 51e^{4\tilde{J}_q} + 32e^{2\tilde{J}_q} + 12)/4(e^{4\tilde{J}_q} + 3e^{2\tilde{J}_q} + 2)^2. \tag{6.49}$$

For $\tilde{J}_q < \ln(2)/2$ *and* $\tilde{J}_q \neq \ln(2)/4$ we have $\coth(\tilde{J}_q) - 3 \neq 0$ and $|\nu_{1,2,3}(\tilde{J}_q)| < \infty$. Hence, in this regime $f(0, \tilde{J}_q) \neq 0$, and therefore we can use the *Lagrange inversion theorem* as in the previous section. Plugging $f(\bar{m}, \tilde{J}_q)$ given by Eq. (6.45) into Eq. (6.42), and using Eq. (6.44) to express $\bar{m}_0$ in terms of $\tilde{J}_0$, we obtain the power series solution given by Eq. (6.7). For completeness, we list the first three non-zero coefficients

$$\begin{aligned}
&\alpha_1(\tilde{J}_0, \tilde{J}_q) = \bar{m}_0 f(0, \tilde{J}_q)/f(\bar{m}_0, \tilde{J}_q),\\
&\alpha_3(\tilde{J}_0, \tilde{J}_q) = \alpha_1^3(\tilde{J}_0, \tilde{J}_q) e^{-4\tilde{J}_q}(4 - e^{2\tilde{J}_q})(1 - e^{2\tilde{J}_q})^2/(4(2 - e^{2\tilde{J}_q})),\\
&\alpha_5(\tilde{J}_0, \tilde{J}_q) = \alpha_1^5(\tilde{J}_0, \tilde{J}_q) \frac{111\cosh(\tilde{J}_q) - 87\cosh(3\tilde{J}_q) - 313\sinh(\tilde{J}_q) + 113\sinh(3\tilde{J}_q)}{8e^{4\tilde{J}_q}(\coth(\tilde{J}_q) - 3)^2} \sinh(\tilde{J}_q).
\end{aligned} \tag{6.50}$$

Note that only terms of $\bar{m}^2$ and $\bar{m}^4$ enter in $f(\bar{m}, \tilde{J}_q)$ given by Eq. (6.45). Therefore, $f(\bar{m}, \tilde{J}_q) = f(-\bar{m}, \tilde{J}_q)$, which implies that $\alpha_{2k} = 0$, $\forall k \in \mathbb{N}$. Furthermore, we also have $\alpha_1(\tilde{J}_0, 0) = 1$ and $\alpha_k(\tilde{J}_0, 0) = 0$, $\forall k \in \{2, 3, \ldots\}$, as in the previous section. The first two coefficients $\alpha_{1,3}$ are displayed in the inset of Fig. 6.2d.

Regime 2. $\tilde{J}_q = \ln(2)/4$

For $\tilde{J}_q = \ln(2)/4$ the outcome of the integral in Eq. (6.35) simplifies drastically, and the resulting expression for the auxiliary function $f(\bar{m}, \ln(2)/4)$ reads

$$\begin{aligned}
&f(\bar{m}, \ln(2)/4)\\
&= \exp\left(\frac{c_1 + c_2\bar{m}^2 - (c_1 + (c_1/4 + c_2)\bar{m}^2 - \sqrt{2}c_3\bar{m}^4)[1 - \bar{m}^2/2]^{\frac{1}{2}}}{\bar{m}^4}\right)(2 + [4 - 2\bar{m}^2]^{\frac{1}{2}})^{c_4},
\end{aligned} \tag{6.51}$$

with the numerical coefficients given by

$$c_1 = 560\sqrt{2} - 792,\ c_2 = 1092 - 772\sqrt{2},\ c_3 = 8(7 - 5\sqrt{2}),\ c_4 = 329 - 232\sqrt{2}.$$

This function $f(\bar{m}, \ln(2)/4)$ attains the following value at $\bar{m} = 0$

$$f(0, \ln(2)/4) = 4^{c_4} \exp\left(\frac{3c_1}{32} + \frac{c_2}{4} + \sqrt{2}c_3\right). \tag{6.52}$$

Hence, $f(0, \ln(2)/4) \neq 0$, and therefore we can use the *Lagrange inversion theorem*. Inserting Eq. (6.51) into Eq. (6.42), we obtain an expression for the coefficients. The result for the first three non-zero coefficients reads

$$\begin{aligned}
\alpha_1(\tilde{J}_0, \ln(2)/4) &= \bar{m}_0 f(0, \ln(2)/4)/f(\bar{m}_0, \ln(2)/4),\\
\alpha_3(\tilde{J}_0, \ln(2)/4) &= \alpha_1^3(\tilde{J}_0, \ln(2)/4)(c_1 + 2c_2 - 8(2\sqrt{2}c_3 + c_4))/4^3,\\
\alpha_5(\tilde{J}_0, \ln(2)/4) &= \alpha_1^5(\tilde{J}_0, \ln(2)/4)\mathcal{A}(c_1, c_2, c_3, c_4)/2^{13},
\end{aligned} \tag{6.53}$$

with

$$\begin{aligned}
\mathcal{A}(c_1,c_2,c_3,c_4) &= 5c_1^2 + 20c_2^2 + 4c_1(9 + 5c_2 - 40\sqrt{2}c_3 - 20c_4)\\
&+ 32c_2(2 - 10\sqrt{2}c_3 - 5c_4) + 64(40c_3^2 + c_4(5c_4 - 3) + 4\sqrt{2}c_3(5c_4 - 1)).
\end{aligned} \tag{6.54}$$

Also here we find that only terms of $\bar{m}^2$ and $\bar{m}^4$ enter in Eq. (6.51), which implies that $\alpha_{2k} = 0,\ \forall k \in \mathbb{N}$. Notably, the coefficients in Eq. (6.50) approach Eq. (6.53) in the neighborhood of $\tilde{J}_q = \ln(2)/4$.

Regime 3. $\tilde{J}_q > \ln(2)/2$

Finally, we focus on a quench in the two-phase domain with $\tilde{J}_q > \ln(2)/2$. Formally, the integral given by Eq. (6.34) does not change w.r.t. the analysis for $\tilde{J}_q < \ln(2)/2$. However, there is a difference in applying the *Lagrange inversion theorem*, since the steady-state magnetization $\bar{m}_\infty(\tilde{J}_q) = \pm e^{2\tilde{J}_q}(e^{4\tilde{J}_q} - 4)^{1/2}/(e^{4\tilde{J}_q} - 2)$ maintains a non-zero value for $\tilde{J}_q > \ln(2)/2$. The relaxation rate now reads $1/\tau_r(\tilde{J}_q) = 4w_{\mathrm{BG}}^{\pm}(\bar{m}_\infty, \tilde{J}_q)\mathfrak{f}''_{\mathrm{BG}}(\bar{m}_\infty, \tilde{J}_q) = (e^{4\tilde{J}_q} - 2)(e^{2\tilde{J}_q} - 2)(e^{2\tilde{J}_q} + 2)^3/(e^{4\tilde{J}_q} + 1)^4$ (see blue line in Fig. 6.2f, right side of the critical coupling). After some algebraic manipulation, we obtain for Eq. (6.34)

$$-\tau_r(\tilde{J}_q)\ln\left(\frac{\bar{m} - \bar{m}_\infty}{f(\bar{m}, \tilde{J}_q)}\right) = t + C, \tag{6.55}$$

where $C = C(\tilde{J}_0, \tilde{J}_q)$ is the integration constant determined by the initial condition. The function $f(\bar{m}, \tilde{J}_q)$ reads

$$f(\bar{m}, \tilde{J}_q) = (e^{4\tilde{J}_q} - 2)^{-1} \prod_{i=1}^{5} f_i(\bar{m}, \tilde{J}_q), \tag{6.56}$$

which we have further divided into the following sub-auxiliary functions

$$\begin{aligned}
f_1(\bar{m}, \tilde{J}_q) &= \exp\left(\frac{2\alpha_-(\bar{m}, \tilde{J}_q)\nu_1(\tilde{J}_q)(2 - e^{2\tilde{J}_q})e^{6\tilde{J}_q}}{\bar{m}^4(1 + e^{2\tilde{J}_q})^4(e^{2\tilde{J}_q} + 2)}\right),\\
f_2(\bar{m}, \tilde{J}_q) &= \exp\left(\frac{\nu_2(\tilde{J}_q)[14 + 20e^{2\tilde{J}_q} + 6e^{4\tilde{J}_q} - e^{-4\tilde{J}_q}(e^{2\tilde{J}_q} + 1)(2 + 13e^{2\tilde{J}_q} + 5e^{4\tilde{J}_q})(\alpha_+ - e^{2\tilde{J}_q})]}{32(1 + e^{2\tilde{J}_q})^4\bar{m}^2}\right),\\
f_3(\bar{m}, \tilde{J}_q) &= [\bar{m}^2(e^{4\tilde{J}_q} - 2)^2(1 - e^{4\tilde{J}_q}) + 4e^{4\tilde{J}_q}(1 - \alpha_+ + e^{2\tilde{J}_q}) - e^{8\tilde{J}_q}(3 - 2\alpha_+ + 2e^{2\tilde{J}_q}) + e^{12\tilde{J}_q}]^{\frac{1}{2}},\\
f_4(\bar{m}, \tilde{J}_q) &= |\bar{m}(e^{4\tilde{J}_q} - 2) + e^{2\tilde{J}_q}(e^{4\tilde{J}_q} - 4)^{1/2}|^{-1},\\
f_5(\bar{m}, \tilde{J}_q) &= \bar{m}^{\nu_1(\tilde{J}_q)},\\
f_6(\bar{m}, \tilde{J}_q) &= \alpha_+(\bar{m}, \tilde{J}_q)^{-\nu_3(\tilde{J}_q)}.
\end{aligned} \tag{6.57}$$

The function $\alpha_\pm(\bar{m}, \tilde{J}_q)$ is given by Eq. (6.37), and furthermore we have introduced

$$\begin{aligned}
\nu_1(\tilde{J}_q) &\equiv e^{-2\tilde{J}_q}(e^{2\tilde{J}_q} + 2)^3(e^{4\tilde{J}_q} - 2)/32,\\
\nu_2(\tilde{J}_q) &\equiv e^{4\tilde{J}_q}(8 - 6e^{4\tilde{J}_q} + e^{8\tilde{J}_q}),\\
\nu_3(\tilde{J}_q) &\equiv e^{\tilde{J}_q}(e^{4\tilde{J}_q} - 2)\,\mathrm{sech}\,(\tilde{J}_q)(28 + 8e^{-4\tilde{J}_q} + 36e^{-2\tilde{J}_q} - 5e^{2\tilde{J}_q} - 3e^{4\tilde{J}_q} - 6\tanh(\tilde{J}_q))/64.
\end{aligned}$$

In order to apply the *Lagrange inversion theorem*, we need to evaluate $f(\bar{m}, \tilde{J}_q)$ at the steady state $\bar{m}_\infty$, which yields

$$f(\bar{m}_\infty, \tilde{J}_q) = (\bar{m}_\infty)^{\nu_1(\tilde{J}_q)}(1 + e^{2\tilde{J}_q} + 2(e^{4\tilde{J}_q} - 2)^{-1})^{-\nu_3(\tilde{J}_q)}e^{2\tilde{J}_q - \nu_4(\tilde{J}_q)}(e^{4\tilde{J}_q} - 2)^{-1/2}, \tag{6.58}$$

where we have defined the auxiliary function

$$\nu_4(\tilde{J}_q) = e^{3\tilde{J}_q}[13 + 8\cosh(2\tilde{J}_q)][\cosh(\tilde{J}_q) - 3\sinh(\tilde{J}_q)][\cosh(\tilde{J}_q) - \sinh(\tilde{J}_q)(6 - \tanh(\tilde{J}_q))]^2.$$

For $\tilde{J}_q > \ln(2)/2$ we find that $f(\bar{m}_\infty, \tilde{J}_q) \neq 0$, and therefore we can apply the *Lagrange inversion theorem*. Upon inverting Eq. (6.55), the final result reads

$$\begin{aligned}
\bar{m}(t, \bar{m}_0, \tilde{J}_q) &= \bar{m}_\infty + \sum_{k=1}^{\infty} \frac{(\bar{m}_0 - \bar{m}_\infty)^k}{f(\bar{m}_0, \tilde{J}_q)^k k!}\left[\frac{d^{k-1} f(\bar{m}, \tilde{J}_q)^k}{dw^{k-1}}\right]_{\bar{m}=\bar{m}_\infty} e^{-kt/\tau_r(\tilde{J}_q)}\\
&= \bar{m}_\infty + \sum_{k=1}^{\infty} \alpha_k(\tilde{J}_0, \tilde{J}_q)e^{-kt/\tau_r(\tilde{J}_q)}.
\end{aligned} \tag{6.59}$$

For completeness, we list the first three non-zero coefficients

$$\begin{aligned}
\alpha_1(\tilde{J}_0, \tilde{J}_q) &= (\bar{m}_0 - \bar{m}_\infty) f(\bar{m}_\infty, \tilde{J}_q)/f(\bar{m}_0, \tilde{J}_q),\\
\alpha_2(\tilde{J}_0, \tilde{J}_q) &= \alpha_1^2 \mathrm{e}^{-6\tilde{J}_q} (\mathrm{e}^{4\tilde{J}_q} - 2)^2 (4\mathrm{e}^{3\tilde{J}_q} \sinh(\tilde{J}_q) - 1)/(\mathrm{e}^{4\tilde{J}_q} - 4)^{1/2},\\
\alpha_3(\tilde{J}_0, \tilde{J}_q) &= \alpha_1^3 \mathrm{e}^{-6\tilde{J}_q} (\mathrm{e}^{4\tilde{J}_q} - 2)^3 \frac{52 - 10\mathrm{e}^{-6\tilde{J}_q} - 24\mathrm{e}^{-4\tilde{J}_q} + 25\mathrm{e}^{-2\tilde{J}_q} - 35\mathrm{e}^{2\tilde{J}_q} - 18\mathrm{e}^{4\tilde{J}_q} + 11\mathrm{e}^{6\tilde{J}_q}}{2(\mathrm{e}^{4\tilde{J}_q} - 4)},
\end{aligned} \tag{6.60}$$

where for convenience we have suppressed the arguments of $\alpha_1(\tilde{J}_0, \tilde{J}_q)$ in the last two equations. The inset of Fig. 6.2e displays the first two coefficients $\alpha_{1,2}$. This concludes our derivation for the relaxation dynamics of the rate function minima.

Appendix 6: Relative Entropy

Here we derive the coefficients γ_k for the power series expansion of the relative entropy per spin, given by Eq. (6.9). The relative entropy is evaluated with the saddle point approximation in the thermodynamic limit, which results in

$$\mathscr{D}_t = \lim_{N\to\infty} \int_{-1}^{1} \mathrm{e}^{-NV(m,\tilde{J}_q,t)} [V_{\mathrm{eq}}(m, \tilde{J}_q) - V(m, \tilde{J}_q, t)] dm \simeq V_{\mathrm{eq}}(\bar{m}, \tilde{J}_q) = \sum_{k=2}^{\infty} \gamma_k(\tilde{J}_0, \tilde{J}_q) \mathrm{e}^{-kt/\tau_r}. \tag{6.61}$$

To arrive at the second equality we have applied the saddle point approximation around the minimum $\bar{m} = \bar{m}(t, \tilde{J}_0, \tilde{J}_q)$ of the rate function $V(m, \tilde{J}_q, t)$ at time t. Note that $V(\bar{m}, \tilde{J}_q, t) = 0$, and therefore only the equilibrium potential $V_{\mathrm{eq}}(\bar{m}, \tilde{J}_q)$ remains after the saddle point approximation. For the final equality we carried out a Taylor expansion around the steady state $\bar{m}_\infty$, and used the power series expansion of $\bar{m}(t, \tilde{J}_0, \tilde{J}_q)$, which is analyzed in Appendix 5. The first three non-zero coefficients in Eq. (6.61) are given by

$$\begin{aligned}
\gamma_2(\tilde{J}_0, \tilde{J}_q) &= \alpha_1^2 V''_{\mathrm{eq}}(\bar{m}_\infty, \tilde{J}_q)/2,\\
\gamma_3(\tilde{J}_0, \tilde{J}_q) &= \alpha_1\alpha_2 V''_{\mathrm{eq}}(\bar{m}_\infty, \tilde{J}_q) + \alpha_1^3 V'''_{\mathrm{eq}}(\bar{m}_\infty, \tilde{J}_q)/6,\\
\gamma_4(\tilde{J}_0, \tilde{J}_q) &= (\alpha_2^2/2 + \alpha_1\alpha_3) V''_{\mathrm{eq}}(\bar{m}_\infty, \tilde{J}_q) + \alpha_1^2\alpha_2 V'''_{\mathrm{eq}}(\bar{m}_\infty, \tilde{J}_q)/2 + \alpha_1^4 V''''_{\mathrm{eq}}(\bar{m}_\infty, \tilde{J}_q)/24,
\end{aligned}$$

where the coefficients $\alpha_i = \alpha_i(\tilde{J}_0, \tilde{J}_q)$ are given by Eqs. (6.50) and (6.60) for quenches in the one- and two-phase domain, respectively. For quenches in the one-phase domain we have $\gamma_3(\tilde{J}_0, \tilde{J}_q) = 0$ since $\bar{m}_\infty = 0$ and $\alpha_2 = V'''_{\mathrm{eq}}(0, \tilde{J}_q) = 0$. The inset of Fig. 6.3a displays the first two non-zero coefficients for quenches in the one- and two-phase domain.

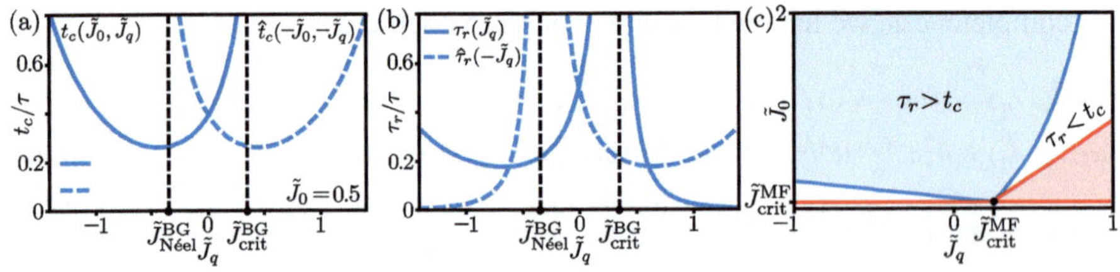

Fig. 6.6 Parity symmetry for the staggered magnetization and the MF dynamical phase diagram. In all panels we consider a lattice with $\bar{z} = 4$. **a, b** Critical time (**a**) and relaxation time (**b**) as a function of $\tilde{J}_q$. The dashed lines correspond to the staggered magnetization dynamics, for which a parity symmetry applies w.r.t. the temperature/coupling $\tilde{J} \to -\tilde{J}$ [see Eq. (6.63)]. **c** Dynamical phase diagram for the MF critical time t_c^{MF} and relaxation τ_r^{MF} time. The red area is forbidden since $\tilde{J}_0 > \tilde{J}_q$. Inside the blue area, the relaxation time is larger than the critical time. The dark blue phase boundary where $t_c^{\mathrm{MF}} = \tau_r^{\mathrm{MF}}$ is given by Eq. (6.66). The MF critical coupling reads $\tilde{J}_{\mathrm{crit}}^{\mathrm{MF}} \equiv 1/\bar{z}$. Figure 6.3b shows the BG dynamical phase diagram

Appendix 7: Staggered Magnetization

Let us define the staggered magnetization $\hat{m} \in [-1, 1]$ in the Ising model as

$$\hat{m} \equiv N^{-1} \sum_{i=1}^{N} (-\sigma_i)^i. \tag{6.62}$$

For perfectly anti-ferromagnetic order we have $\hat{m} = \pm 1$, and for anti-ferromagnetic disorder $\hat{m} = 0$. Based on the works in [94–96] we know that the BG free energy density $\tilde{\mathrm{f}}_{\mathrm{BG}}(m, \tilde{J})$ obeys the following parity symmetry w.r.t. the staggered magnetization

$$\tilde{\mathrm{f}}_{\mathrm{BG}}(m, \tilde{J}) = \tilde{\mathrm{f}}_{\mathrm{BG}}(\hat{m}, -\tilde{J}). \tag{6.63}$$

Therefore, our results for the critical time, relaxation time, and dynamical phase diagram also apply for dynamics of staggered magnetization upon inverting the temperature/coupling $\tilde{J} \to -\tilde{J}$. In Fig. 6.6a, b we depict the critical time $\hat{t}_c$ (a) and relaxation time $\hat{\tau}_r$ (b) for the dynamics of the staggered magnetization with the blue dashed lines.

Appendix 8: MF Dynamical Phase Diagram

Fig. 6.6c depicts the MF dynamical phase diagram. To obtain the blue shaded area where $\tau_r^{\mathrm{MF}} > t_c^{\mathrm{MF}}$, we first compute the MF critical time. Inserting the MF transition rates and free energy density into Eq. (6.4), we obtain the MF critical time

$$t_c^{\mathrm{MF}}(\tilde{J}_0, \tilde{J}_q) = \frac{1}{4(1 - \bar{z}J_q)} \ln \left(\frac{\bar{z}\tilde{J}_q - \bar{z}\tilde{J}_0}{1 - \bar{z}\tilde{J}_0} \right), \tag{6.64}$$

which is also reported in [1, 3, 4] for $\bar{z} = 1$. The MF relaxation time is given by $\tau_r^{\mathrm{MF}}(\tilde{J}_q) \equiv 1/4w_{\mathrm{MF}}^{\pm}(\bar{m}, \tilde{J}_q)\tilde{\mathrm{f}}''_{\mathrm{MF}}(\bar{m}, \tilde{J}_q)$, where $\bar{m} = \arg\min_m \tilde{\mathrm{f}}_{\mathrm{MF}}(m, \tilde{J}_q)$ is given by the transcendental equation

$$\bar{m} = \tanh(\bar{z}\tilde{J}_q\bar{m}). \tag{6.65}$$

Equating t_c^{MF} and τ_r^{MF}, we obtain the dark blue boundary line

$$\bar{z}\tilde{J}_0^{\dagger} = \frac{\bar{z}\tilde{J}_q \exp\left(\frac{2(1+\bar{m})(\bar{z}\tilde{J}_q - 1)}{1-(1-\bar{m}^2)\bar{z}\tilde{J}_q} \mathrm{e}^{-\bar{z}\bar{m}\tilde{J}_q}\right) - 1}{\exp\left(\frac{2(1+\bar{m})(\bar{z}\tilde{J}_q - 1)}{1-(1-\bar{m}^2)\bar{z}\tilde{J}_q} \mathrm{e}^{-\bar{z}\bar{m}\tilde{J}_q}\right) - 1}. \tag{6.66}$$

For $\tilde{J}_0 > \tilde{J}_0^{\dagger}$ (blue region) the MF relaxation time is larger than the critical time, i.e. $\tau_r^{\mathrm{MF}} > t_c^{\mathrm{MF}}$. For $1/\bar{z} < \tilde{J}_0 < \tilde{J}_0^{\dagger}$ (white region) the MF critical time is larger than the relaxation time.

References

1. Meibohm J, Esposito M (2022) Finite-time dynamical phase transition in nonequilibrium relaxation. Phys Rev Lett 128:110603. [Online]. Available: https://link.aps.org/doi/10.1103/PhysRevLett.128.110603
2. Meibohm J, Esposito M (2022) Landau theory for finite-time dynamical phase transitions. arXiv preprint arXiv:2205.10311
3. Külske C, Le Ny A (2007) Spin-flip dynamics of the curie-weiss model: Loss of gibbsianness with possibly broken symmetry. Commun Math Phys 271(2):431–454. [Online]. Available: http://dx.doi.org/10.1007/s00220-007-0201-y
4. Ermolaev V, Külske C (2010) Low-temperature dynamics of the curie-weiss model: periodic orbits, multiple histories, and loss of gibbsianness. J Stat Phys 141(5):727–756. [Online]. Available: http://dx.doi.org/10.1007/s10955-010-0074-x
5. Blom K, Godec A (2022) Global speed limit for finite-time dynamical phase transition in nonequilibrium relaxation. [Online]. Available: https://arxiv.org/abs/2209.14287
6. Dattagupta S (2012) Relaxation phenomena in condensed matter physics. Elsevier
7. Wolfgang Haase SW (2003) Relaxation phenomena. Springer Berlin Heidelberg. [Online]. Available: http://dx.doi.org/10.1007/978-3-662-09747-2
8. Onsager L (1931) Phys Rev 37(4):405–426. [Online]. Available: https://link.aps.org/doi/10.1103/PhysRev.37.405
9. Onsager L (1931) Phys Rev 38(12):2265–2279. [Online]. Available: https://link.aps.org/doi/10.1103/PhysRev.38.2265
10. Kubo R, Yokota M, Nakajima S (1957) Statistical-mechanical theory of irreversible processes. ii. response to thermal disturbance. J Phys Soc Jpn 12(11):1203–1211. [Online]. Available: http://dx.doi.org/10.1143/JPSJ.12.1203
11. Seifert U, Speck T (2010) Fluctuation-dissipation theorem in nonequilibrium steady states. EPL (Europhys Lett) 89(1):10007. [Online]. Available: https://doi.org/10.1209/0295-5075/89/10007

12. Baiesi M, Maes C (2013) New J Phys 15(1): 013004. [Online]. Available: http://dx.doi.org/10.1088/1367-2630/15/1/013004
13. Wu W, Wang J (2020) Generalized fluctuation-dissipation theorem for non-equilibrium spatially extended systems. Front Phys 8. [Online]. Available: http://dx.doi.org/10.3389/fphy.2020.567523
14. Cugliandolo LF, Dean DS, Kurchan J (1997) Fluctuation-dissipation theorems and entropy production in relaxational systems Phys Rev Lett 79:2168–2171. [Online]. Available: https://link.aps.org/doi/10.1103/PhysRevLett.79.2168
15. Lippiello E, Baiesi M, Sarracino A (2014) Nonequilibrium fluctuation-dissipation theorem and heat production. Phys Rev Lett 112:140602. [Online]. Available: https://link.aps.org/doi/10.1103/PhysRevLett.112.140602
16. Maes C (2020) Frenesy: time-symmetric dynamical activity in nonequilibria. Phys Rep 850:1–33. [Online]. Available: https://www.sciencedirect.com/science/article/pii/S0370157320300120
17. Lu Z, Raz O (2017) Proc Natl Acad Sci 114(20):5083–5088. [Online]. Available: http://www.pnas.org/lookup/doi/10.1073/pnas.1701264114
18. Klich I, Raz O, Hirschberg O et al (2019) Phys Rev X 9(2):021060. [Online]. Available: https://link.aps.org/doi/10.1103/PhysRevX.9.021060
19. Lasanta A, Vega Reyes F, Prados A et al (2017) When the hotter cools more quickly: Mpemba effect in granular fluids. Phys Rev Lett 119:148001. [Online]. Available: https://link.aps.org/doi/10.1103/PhysRevLett.119.148001
20. Busiello DM, Gupta D, Maritan A (2021) Inducing and optimizing markovian mpemba effect with stochastic reset. New J Phys 23(10):103012. [Online]. Available: http://dx.doi.org/10.1088/1367-2630/ac2922
21. Gal A, Raz O (2020) Precooling strategy allows exponentially faster heating. Phys Rev Lett 124:060602. [Online]. Available: https://link.aps.org/doi/10.1103/PhysRevLett.124.060602
22. Zulkowski PR, DeWeese MR (2015) Optimal control of overdamped systems. Phys Rev E 92:032117. [Online]. Available: https://link.aps.org/doi/10.1103/PhysRevE.92.032117
23. Frim AG, Zhong A, Chen S-F et al (2021) Engineered swift equilibration for arbitrary geometries. Phys Rev E 103:L030102. [Online]. Available: https://link.aps.org/doi/10.1103/PhysRevE.103.L030102
24. Lapolla A, Godec A (2020) Faster uphill relaxation in thermodynamically equidistant temperature quenches. Phys Rev Lett 125:110602. [Online]. Available: https://link.aps.org/doi/10.1103/PhysRevLett.125.110602
25. Meibohm J, Forastiere D, Adeleke-Larodo T et al (2021) Relaxation-speed crossover in anharmonic potentials. Phys Rev E 104:L032105. [Online]. Available: https://link.aps.org/doi/10.1103/PhysRevE.104.L032105
26. Manikandan SK (2021) Equidistant quenches in few-level quantum systems. Phys Rev Res 3:043108. [Online]. Available: https://link.aps.org/doi/10.1103/PhysRevResearch.3.043108
27. Van Vu T, Hasegawa Y (2021) Toward relaxation asymmetry: Heating is faster than cooling. Phys Rev Res 3:043160. [Online]. Available: https://link.aps.org/doi/10.1103/PhysRevResearch.3.043160
28. Graham R, Tél T (1984) On the weak-noise limit of fokker-planck models. J Stat Phys 35(5–6):729–748. [Online]. Available: http://dx.doi.org/10.1007/BF01010830
29. Graham R, Tél T (1985) Weak-noise limit of fokker-planck models and nondifferentiable potentials for dissipative dynamical systems. Phys Rev A 31:1109–1122. [Online]. Available: https://link.aps.org/doi/10.1103/PhysRevA.31.1109
30. Bouchet F, Gawedzki K, Nardini C (2016) Perturbative calculation of quasi-potential in non-equilibrium diffusions: A mean-field example. J Stat Phys 163(5):1157–1210. [Online]. Available: http://dx.doi.org/10.1007/s10955-016-1503-2
31. Bertini L, De Sole A, Gabrielli D et al (2001) Fluctuations in stationary nonequilibrium states of irreversible processes. Phys Rev Lett 87:040601. [Online]. Available: https://link.aps.org/doi/10.1103/PhysRevLett.87.040601

32. Bertini L, Sole AD, Gabrielli D et al (2010) Lagrangian phase transitions in nonequilibrium thermodynamic systems. J Stat Mech 2010(11):L11001. [Online]. Available: https://doi.org/10.1088/1742-5468/2010/11/l11001
33. Bunin G, Kafri Y, Podolsky D (2012) Non-differentiable large-deviation functionals in boundary-driven diffusive systems. J Stat Mech 2012(10):L10001. [Online]. Available: http://dx.doi.org/10.1088/1742-5468/2012/10/L10001
34. Bunin G, Kafri Y, Podolsky D (2013) Cusp singularities in boundary-driven diffusive systems. J Stat Phys 152(1):112–135. [Online]. Available: http://dx.doi.org/10.1007/s10955-013-0752-6
35. Baek Y, Kafri Y (2015) Singularities in large deviation functions. J Stat Mech 2015(8):P08026. [Online]. Available: http://dx.doi.org/10.1088/1742-5468/2015/08/P08026
36. Garrahan JP, Jack RL, Lecomte V et al (2007) Dynamical first-order phase transition in kinetically constrained models of glasses. Phys Rev Lett 98:195702. [Online]. Available: https://link.aps.org/doi/10.1103/PhysRevLett.98.195702
37. Garrahan JP, Jack RL, Lecomte V et al (2009) First-order dynamical phase transition in models of glasses: an approach based on ensembles of histories. J Phys A Math Theor 42(7):075007. [Online]. Available: http://dx.doi.org/10.1088/1751-8113/42/7/075007
38. Chandler D, Garrahan JP (2010) Dynamics on the way to forming glass: Bubbles in space-time. Annu Rev Phys Chem 61(1):191–217. [Online]. Available: https://doi.org/10.1146/annurev.physchem.040808.090405
39. Garrahan JP, Lesanovsky I (2010) Thermodynamics of quantum jump trajectories. Phys Rev Lett 104:160601. [Online]. Available: https://link.aps.org/doi/10.1103/PhysRevLett.104.160601
40. Ates C, Olmos B, Garrahan JP et al (2012) Dynamical phases and intermittency of the dissipative quantum ising model. Phys Rev A 85:043620. [Online]. Available: https://link.aps.org/doi/10.1103/PhysRevA.85.043620
41. Hickey JM, Flindt C, Garrahan JP (2014) Intermittency and dynamical lee-yang zeros of open quantum systems. Phys Rev E 90:062128. [Online]. Available: https://link.aps.org/doi/10.1103/PhysRevE.90.062128
42. Jack RL, Sollich P (2013) Large deviations of the dynamical activity in the east model: analysing structure in biased trajectories. J Phys A Math Theor 47(1):015003. [Online]. Available: http://dx.doi.org/10.1088/1751-8113/47/1/015003
43. Gorissen M, Lazarescu A, Mallick K et al (2012) Exact current statistics of the asymmetric simple exclusion process with open boundaries. Phys Rev Lett 109:170601. [Online]. Available: https://link.aps.org/doi/10.1103/PhysRevLett.109.170601
44. Espigares CP, Garrido PL, Hurtado PI (2013) Dynamical phase transition for current statistics in a simple driven diffusive system. Phys Rev E 87:032115. [Online]. Available: https://link.aps.org/doi/10.1103/PhysRevE.87.032115
45. Tsobgni Nyawo P, Touchette H (2016) Large deviations of the current for driven periodic diffusions. Phys Rev E 94:032101. [Online]. Available: https://link.aps.org/doi/10.1103/PhysRevE.94.032101
46. Tizón-Escamilla N, Pérez-Espigares C, Garrido PL et al (2017) Order and symmetry breaking in the fluctuations of driven systems. Phys Rev Lett 119:090602. [Online]. Available: https://link.aps.org/doi/10.1103/PhysRevLett.119.090602
47. Mehl J, Speck T, Seifert U (2008) Large deviation function for entropy production in driven one-dimensional systems. Phys Rev E 78:011123. [Online]. Available: https://link.aps.org/doi/10.1103/PhysRevE.78.011123
48. Speck T, Engel A, Seifert U (2012) The large deviation function for entropy production: the optimal trajectory and the role of fluctuations. J Stat Mech 2012(12):P12001. [Online]. Available: http://dx.doi.org/10.1088/1742-5468/2012/12/P12001
49. Tsobgni Nyawo P, Touchette H (2016) A minimal model of dynamical phase transition. EPL (Europhys Lett) 116(5):50009. [Online]. Available: http://dx.doi.org/10.1209/0295-5075/116/50009

50. Jack RL, Thompson IR, Sollich P (2015) Hyperuniformity and phase separation in biased ensembles of trajectories for diffusive systems. Phys Rev Lett 114:060601. [Online]. Available: https://link.aps.org/doi/10.1103/PhysRevLett.114.060601
51. Harris RJ, Touchette H (2017) Phase transitions in large deviations of reset processes. J Phys A Math Theor 50(10):10LT01. [Online]. Available: http://dx.doi.org/10.1088/1751-8121/aa5734
52. Barratt F, Comas AB, Crowley P et al (2021) Dissipative failure of adiabatic quantum transport as a dynamical phase transition. Phys Rev A 103:052427. [Online]. Available: https://link.aps.org/doi/10.1103/PhysRevA.103.052427
53. Pietzonka P, Ritort F, Seifert U (2017) Finite-time generalization of the thermodynamic uncertainty relation. Phys Rev E 96:012101
54. Dechant A (2018) Multidimensional thermodynamic uncertainty relations. J Phys A Math Theor 52(3):035001
55. Liu K, Gong Z, Ueda M (2020) Thermodynamic uncertainty relation for arbitrary initial states. Phys Rev Lett 125:140602. [Online]. Available: https://link.aps.org/doi/10.1103/PhysRevLett.125.140602
56. Koyuk T, Seifert U (2019) Operationally accessible bounds on fluctuations and entropy production in periodically driven systems. Phys Rev Lett 122:23060230601
57. Koyuk T, Seifert U (2020) Thermodynamic uncertainty relation for time-dependent driving. Phys Rev Lett 125(26):260604
58. Dieball C, Godec A (2022) Direct route to thermodynamic uncertainty relations
59. Mandelstam L, Tamm I (1945) The uncertainty relation between energy and time in nonrelativistic quantum mechanics. J Phys USSR 9
60. Bhattacharyya K (1983) Quantum decay and the mandelstam-tamm-energy inequality. J Phys A Math Gen 16(13):2993–2996. [Online]. Available: http://dx.doi.org/10.1088/0305-4470/16/13/021
61. Anandan J, Aharonov Y (1990) Geometry of quantum evolution. Phys Rev Lett 65:1697–1700. [Online]. Available: https://link.aps.org/doi/10.1103/PhysRevLett.65.1697
62. Pfeifer P (1993) How fast can a quantum state change with time? Phys Rev Lett 70:3365–3368. [Online]. Available: https://link.aps.org/doi/10.1103/PhysRevLett.70.3365
63. Margolus N, Levitin LB (1998) The maximum speed of dynamical evolution. Phys D 120(1):188–195. [Online]. Available: https://www.sciencedirect.com/science/article/pii/S0167278998000542
64. Lloyd S (2000) Ultimate physical limits to computation. Nature 406(6799):1047–1054. [Online]. Available: http://dx.doi.org/10.1038/35023282
65. Giovannetti V, Lloyd S, Maccone L (2003) Quantum limits to dynamical evolution. Phys Rev A 67:052109. [Online]. Available: https://link.aps.org/doi/10.1103/PhysRevA.67.052109
66. Deffner S, Lutz E (2013) Energy–time uncertainty relation for driven quantum systems. J Phys A Math Theor 46(33):335302. [Online]. Available: http://dx.doi.org/10.1088/1751-8113/46/33/335302
67. Taddei MM, Escher BM, Davidovich L et al (2013) Quantum speed limit for physical processes. Phys Rev Lett 110:050402. [Online]. Available: https://link.aps.org/doi/10.1103/PhysRevLett.110.050402
68. del Campo A, Egusquiza IL, Plenio MB et al (2013) Quantum speed limits in open system dynamics. Phys Rev Lett 110:050403. [Online]. Available: https://link.aps.org/doi/10.1103/PhysRevLett.110.050403
69. Deffner S, Lutz E (2013) Quantum speed limit for non-markovian dynamics. Phys Rev Lett 111:010402. [Online]. Available: https://link.aps.org/doi/10.1103/PhysRevLett.111.010402
70. García-Pintos LP, Nicholson SB, Green JR et al (2022) Unifying quantum and classical speed limits on observables. Phys Rev X 12:011038. [Online]. Available: https://link.aps.org/doi/10.1103/PhysRevX.12.011038
71. Okuyama M, Ohzeki M (2018) Quantum speed limit is not quantum. Phys Rev Lett 120:070402. [Online]. Available: https://link.aps.org/doi/10.1103/PhysRevLett.120.070402

72. Shanahan B, Chenu A, Margolus N et al (2018) Quantum speed limits across the quantum-to-classical transition. Phys Rev Lett 120:070401. [Online]. Available: https://link.aps.org/doi/10.1103/PhysRevLett.120.070401
73. Shiraishi N, Funo K, Saito K (2018) Speed limit for classical stochastic processes. Phys Rev Lett 121:070601. [Online]. Available: https://link.aps.org/doi/10.1103/PhysRevLett.121.070601
74. Aurell E, Mejía-Monasterio C, Muratore-Ginanneschi P (2011) Optimal protocols and optimal transport in stochastic thermodynamics. Phys Rev Lett 106:250601. [Online]. Available: https://link.aps.org/doi/10.1103/PhysRevLett.106.250601
75. Ito S, Dechant A (2020) Stochastic time evolution, information geometry, and the cramér-rao bound. Phys Rev X 10:021056. [Online]. Available: https://link.aps.org/doi/10.1103/PhysRevX.10.021056
76. Aurell E, Gawedzki K, Mejía-Monasterio C et al (2012) Refined second law of thermodynamics for fast random processes. J Stat Phys 147(3):487–505. [Online]. Available: http://dx.doi.org/10.1007/s10955-012-0478-x
77. Vo VT, Van Vu T, Hasegawa Y (2020) Unified approach to classical speed limit and thermodynamic uncertainty relation. Phys Rev E 102:062132. [Online]. Available: https://link.aps.org/doi/10.1103/PhysRevE.102.062132
78. Falasco G, Esposito M (2020) Dissipation-time uncertainty relation. Phys Rev Lett 125:120604. [Online]. Available: https://link.aps.org/doi/10.1103/PhysRevLett.125.120604
79. Shiraishi N, Saito K (2019) Information-theoretical bound of the irreversibility in thermal relaxation processes. Phys Rev Lett 123:110603. [Online]. Available: https://link.aps.org/doi/10.1103/PhysRevLett.123.110603
80. Yoshimura K, Ito S (2021) Thermodynamic uncertainty relation and thermodynamic speed limit in deterministic chemical reaction networks. Phys Rev Lett 127:160601. [Online]. Available: https://link.aps.org/doi/10.1103/PhysRevLett.127.160601
81. Bray A (1993) Theory of phase ordering kinetics. Phys A 194(1):41–52. [Online]. Available: https://www.sciencedirect.com/science/article/pii/0378437193903385
82. Bethe HA (1935) Statistical theory of superlattices. Proc Math Phys Eng Sci 150(871):552–575
83. Guggenheim EA (1935) The statistical mechanics of regular solutions. Proc Math Phys Eng Sci 148(864):304–312
84. Kawasaki K (1966) Diffusion constants near the critical point for time-dependent ising models. I. Phys Rev 145:224–230. [Online]. Available: https://link.aps.org/doi/10.1103/PhysRev.145.224
85. Kadanoff LP, Swift J (1968) Transport coefficients near the critical point: a master-equation approach. Phys Rev 165:310–322. [Online]. Available: https://link.aps.org/doi/10.1103/PhysRev.165.310
86. Blom K, Godec A (2021) Criticality in cell adhesion. Phys Rev X 11(3):031067. [Online]. Available: https://link.aps.org/doi/10.1103/PhysRevX.11.031067
87. Saito Y, Kubo R (1976) Nonlinearity in cooperative systems-dynamical bethe-ising model. J Stat Phys 15:233–253. [Online]. Available: https://doi.org/10.1007/BF01012879
88. Dykman MI, Mori E, Ross J et al (1994) Large fluctuations and optimal paths in chemical kinetics. J Chem Phys 100(8):5735–5750. [Online]. Available: https://doi.org/10.1063/1.467139
89. Imparato A, Peliti L (2005) Work-probability distribution in systems driven out of equilibrium. Phys Rev E 72:046114. [Online]. Available: https://link.aps.org/doi/10.1103/PhysRevE.72.046114
90. Chatterjee J, Polley D, Pattabi A et al (2022) Rkky exchange bias mediated ultrafast all-optical switching of a ferromagnet. Adv Funct Mater 32(8):2107490. [Online]. Available: https://onlinelibrary.wiley.com/doi/abs/10.1002/adfm.202107490
91. Hakonen P, Lounasmaa OV (1994) Negative absolute temperatures: "hot" spins in spontaneous magnetic order. Science 265(5180):1821–1825. [Online]. Available: http://www.jstor.org/stable/2884649

92. Frenkel D, Warren PB (2015) Gibbs, Boltzmann, and negative temperatures. Am J Phys 83(2):163–170. [Online]. Available: http://dx.doi.org/10.1119/1.4895828
93. Ziman JM (1951) Antiferromagnetism by the Bethe method. Proc Phys Soc A 64(12):1108–1112. [Online]. Available: https://doi.org/10.1088/0370-1298/64/12/307
94. Katsura S, Takizawa M (1974) Bethe lattice and the Bethe approximation. Prog Theor Exp Phys 51(1):82–98. [Online]. Available: https://doi.org/10.1143/PTP.51.82
95. Ono I (1984) Phase transition of ferro- and antiferromagnetic r-state Potts models by the Bethe approximation. J Phys C Solid State Phys 17(20):3615–3631. [Online]. Available: https://doi.org/10.1088/0022-3719/17/20/013
96. Peruggi F, di Liberto F, Monroy G (1983) The potts model on bethe lattices. I. General results. J Phys A Math Theor 16(4):811–827. [Online]. Available: https://doi.org/10.1088/0305-4470/16/4/018
97. Love ER (1980) 64.4 some logarithm inequalities. Math Gaz 64(427):55–57. [Online]. Available: http://www.jstor.org/stable/3615890
98. Squartini T, de Mol J, den Hollander F et al (2015) Breaking of ensemble equivalence in networks. Phys Rev Lett 115:268701. [Online]. Available: https://link.aps.org/doi/10.1103/PhysRevLett.115.268701
99. Touchette H (2015) Equivalence and nonequivalence of ensembles: thermodynamic, macrostate, and measure levels. J Stat Phys 159:987–1016. [Online]. Available: https://doi.org/10.1007/s10955-015-1212-2
100. Lebowitz JL, Bergmann PG (1957) Irreversible gibbsian ensembles. Ann Phys (N Y) 1(1):1–23. [Online]. Available: https://www.sciencedirect.com/science/article/pii/0003491657900027
101. Mackey MC (1989) The dynamic origin of increasing entropy. Rev Mod Phys 61:981–1015. [Online]. Available: https://link.aps.org/doi/10.1103/RevModPhys.61.981
102. Qian H (2013) A decomposition of irreversible diffusion processes without detailed balance. J Math Phys 54(5):053302. [Online]. Available: https://doi.org/10.1063/1.4803847
103. Van den Broeck C, Esposito M (2010) Three faces of the second law. ii. fokker-planck formulation. Phys Rev E 82:011144. [Online]. Available: https://link.aps.org/doi/10.1103/PhysRevE.82.011144
104. Esposito M, Van den Broeck C (2010) Three detailed fluctuation theorems. Phys Rev Lett 104:090601. [Online]. Available: https://link.aps.org/doi/10.1103/PhysRevLett.104.090601
105. Vaikuntanathan S, Jarzynski C (2009) Dissipation and lag in irreversible processes. EPL 87(6):60005. [Online]. Available: https://doi.org/10.1209/0295-5075/87/60005

Chapter 7
Conclusion and Outlook

The characteristic of scientific progress is our knowing that we did not know.

Gaston Bachelard

7.1 Summary and Conclusion

The nearest-neighbor Ising model is one of the most thoroughly studied and applied models in the field of statistical physics (see also Chap. 1). Despite its simplistic formulation of an interacting many-body system, exact analytical results for dimensions greater or equal to two are difficult (and maybe impossible [1]) to obtain, and therefore many questions remain unanswered. Trying to answer these questions is not only relevant in the context of the Ising model, but also gives fundamental insight into the properties of *any* interacting many-body system with sufficiently short-range interactions.[1] It is therefore, that the Ising model has a profound history, and remains hitherto an active research topic.

In this thesis we have studied the equilibrium and kinetic properties of the nearest-neighbor Ising model within the so-called BG approximation, which takes into account pair correlations that arise due to local interactions between nearest-neighbor spins. With the BG approximation one can obtain analytical results for the fixed-magnetization partition function, free energy density, and LEQ transition rates, which are exact on the Bethe lattice (see Figs. 2.1 and 6.1a for examples). Furthermore, the BG approximation provides a more accurate description of the nearest-neighbor Ising model than the well-known MF approximation, which neglects correlations and only takes into account average spin values. Below we provide a brief overview of the main results and conclusions obtained per chapter, and place them into context w.r.t. other related works on the Ising model.

[1] A relevant example is virus spreading through close-contact transmission [2–4].

K. Blom, *Pair-Correlation Effects in Many-Body Systems*, Springer Theses,
https://doi.org/10.1007/978-3-031-29612-3_7

In Chap. 2 we introduced the BG approximation for a uniform nearest-neighbor Ising model, and subsequently derived the fixed-magnetization partition function Eq. (2.13), free energy landscape in Eq. (2.14), and phase diagram shown in Fig. 2.3 for the fraction of down spins φ. Here "*uniform*" means that we neglected spatial fluctuations in φ. For strong coupling we found that the BG approximation is considerably more accurate than the MF approximation w.r.t. exact results for finite system sizes (see Fig. 2.7). Analytical results for the BG free energy density are similarly derived in the original works by Bethe [5] and Guggenheim [6], or can be found in modern textbooks [7].

In Chap. 3 we applied the BG approximation to a non-uniform nearest-neighbor Ising model, allowing for spatial fluctuations of the fraction of down spins, i.e. $\varphi = \varphi(x, y)$. By carrying out a spin block limit (see Fig. 3.1), we derived a Cahn-Hilliard field theory given by Eq. (3.45), with a concentration-dependent gradient energy coefficient in Eq. (3.33). Subsequently, we analyzed the one-dimensional equilibrium concentration profile, and found for strong coupling a widening of the interface (see Fig. 3.2), which is not present in the MF profile (see Fig. 3.3b). Finally, we benchmarked the BG and MF partition functions against exact results for finite systems, and found (similar to Fig. 2.7) that the BG approximation is considerably more accurate (see Fig. 3.4). To the best of our knowledge we are the first to derive a two-dimensional Cahn-Hilliard field theory within the BG approximation. Kikuchi derived similar results for a one-dimensional concentration profile [8], but erroneously refuted the interface widening as incorrect since "*These effects are contrary to physical intuition ...*" (p. 148 in [8]). Presumably, this is due to a lack of physical understanding behind the developed theory. Here, we closed this gap in Chap. 4.

In Chap. 4 we sought for a physical explanation behind the interface widening of the BG one-dimensional concentration profile observed in Chap. 3. Based on a detailed comparison with MC simulations (see Fig. 4.2), and exact results in the strong coupling limit (see Fig. 4.3), we found that interface widening is the result of an entropy-driven interface delocalization transition. The delocalization transition gives rise to translational invariance of the instantaneous interface position for sufficiently strong coupling, which arises due to decreased amplitude of capillary wave fluctuations. The MF approximation does not account for the delocalization transition, and therefore does not show interface widening. Finally, in Sects. 4.6 and 4.7 we analyzed how the delocalization transition affects spinodal decomposition and nucleation, where we found similar non-monotonic behavior w.r.t. the coupling strength (see Figs. 4.3f and 4.4). Our results directly refute the conclusion by Kikuchi in [8], and give a firm physical basis for interface widening in the BG approximation. Hence, our analysis revealed how a Cahn-Hilliard field theory with pair correlations includes the effect of capillary waves.

In Chap. 5 we used the Ising model to construct a mesoscopic model for nearest-neighbor interacting diffusing adhesion bonds in the presence of an external force. The analogy between cell adhesion and the Ising model is based on the works of Refs. [9, 10], where it was shown that membrane fluctuations induce an effective interaction between neighbouring adhesion bonds (see Figs. 1.1b and 5.2). After we analyzed the equilibrium properties in the absence and presence of an external pushing and pulling force (see Sect. 5.3), we analyzed the kinetics of cluster formation and dissolution. By employing the highly-accurate LEQ approximation (see Fig. 5.7), we obtained analytical expression for the MFPT to cluster dissolution and formation in the thermodynamic limit [see Eq. (5.20)]. Here we discovered, for the first time, the existence of a dynamical critical coupling, where the MFPT attains a global minimum *and* undergoes a first-order discontinuity (see Fig. 5.8). From a biological perspective the dynamical critical coupling reflects an optimal bending rigidity of the cell membrane where cluster dissolution and formation are the fastest, and most sensitive to external fluctuations (see Fig. 5.11). Finally, in Sect. 5.5 we hypothesized on the biological implications of the statical and dynamical critical coupling, and placed our results into context with experimental results. Within the context of the Ising model the dynamical critical coupling corresponds to a first-order discontinuity in the mean magnetization reversal time (see Sect. 5.6).

In Chap. 6 we focused on continuous-time relaxation dynamics of the magnetization probability density in the nearest-neighbor Ising model on the Bethe lattice. Inspired by recent works [11–14], which observed a finite-time dynamical phase transition in the relaxation of the Curie-Weiss MF model, we analyzed relaxation dynamics within the BG approximation. Our analysis revealed, for the first time, the existence of the finite-time dynamical phase transition for the nearest-neighbor Ising model on the Bethe lattice (see Figs. 6.1c and 6.2a). Furthermore, within the BG approximation we proved the existence of a speed limit for the (i) critical time at which the dynamical phase transition appears (see Eq. 6.6 and Fig. 6.2c, blue line), and (ii) relaxation time for the average magnetization (see Sect. 6.6 and Fig. 6.2f, blue line). Both speed limits are not present in the MF approximation (see Fig. 6.2c and f, red line), and are therefore a direct result of pair correlations. This is to be expected, as the speed limit is a result of *local* spin configurations that impose kinetic constraints.

7.2 Discussion

Based on the various outcomes, it is clear that pair-correlations have non-trivial effects on the equilibrium and kinetic properties of the nearest-neighbor Ising model. The work shown in this thesis only scratched the surface, and we expect many more surprising outcomes to follow based on the BG approximation.

The applicability of any derived result is limited by the underlying model assumptions, and this is certainly also the case for the BG approximation. Based on Kikuchi's CVA, introduced in Sect. 1.1.7, we know that the BG approximation is not the end of the line. Instead of placing spin pairs onto a given lattice, one could also consider squares or triangles of spins (see Sect. C in [15]). It is not clear how much of the results presented in this thesis can be derived analytically with this procedure, but it is certainly an interesting direction to consider.

Another interesting direction for future studies are irreversible kinetic Ising models with broken DB. Breaking DB comes in many flavours, including asymmetric interaction strengths [16, 17], dynamic constraints [18, 19], competing dynamics at two temperatures [20–23], or through an external energy source [24]. The list of examples presented here is limited, but should give the reader an impression of recent works. Here it would be interesting to consider the non-equilibrium extension of the LEQ approximation, which maps the full dynamics onto a birth-death process. In combination with the BG approximation, this allows us to construct a non-equilibrium free energy density, which one can further analyze in the context of non-equilibrium phase separation [25, 26].

Despite the detailed analysis, various verification techniques, and comparisons with exact results, our physical understanding of the presented outcomes is far from complete. In the end of Chaps. 4–6 we have provided a brief discussion on our derived results. Here we extend the discussion, and provide per chapter a list of unanswered questions

- Chapter 3:
  1. Can we apply the same spin block procedure, in combination with the LEQ approximation, to obtain a Cahn-Hilliard free energy functional for an irreversible Ising model?
- Chapter 4:
  1. We know that the MF approximation does not account for capillary wave fluctuations (see [27, 28]). This would imply that the instantaneous interface remains perfectly stiff, and therefore has translational invariance. Yet, the MF concentration profile is not flat for any coupling value above the critical coupling (see Fig. 4.3f, red line). Is there another mechanism that maintains the MF interface steepness at a nonzero value?
  2. How can we derive a Cahn-Hilliard free energy functional where we fix the instantaneous interface position? Here, we would expect that the interface broadening disappears.
- Chapter 5:
  1. What is the *exact* value of the dynamical critical coupling for the two-dimensional Ising model, where the MFPT for magnetization-reversal attains a minimum? And what is the corresponding speed limit? Which physical principles determine the speed limit?
  2. Does the dynamical critical coupling survive when we consider non-equilibrium kinetics with broken DB?
  3. We assumed that the coupling strength $\tilde{J}$ and intrinsic binding-affinity $\tilde{\mu}$ are constant for each adhesion bond. In a real cellular environment this will not be the case, as external factors can locally stiffen/soften the membrane, or pull on the adhesion bonds. What happens when we consider a local coupling strength $\tilde{J}_{ij}$ (i.e. a spin-glass) and/or binding-affinity $\tilde{\mu}_i$? Can we construct an approximation scheme which incorporates this?

- Chapter 6:

 1. What physical mechanism determines the lower bound of the critical time? We know it arises from local kinetic constraints due to the antiferromagnetic order. But can we quantify this?
 2. Why does the cusp location undergoes a symmetry breaking transition for coupling values above the symmetry-breaking coupling $\tilde{J}_0 > \tilde{J}_{0,\mathrm{BG}}^{\mathrm{SB}}(\tilde{J}_q)$ (see Fig. 6.4)?

The list of questions presented above should give the reader some idea about the many questions that remain unanswered. Hopefully in the future we can provide an answer to some, but preferably all, of these questions.

7.3 Concluding Perspective

Finally, let me conclude with a personal reflection on the last 4 years. When I started my Ph.D., I had a rather naive idea about what it means to understand something. At first, I thought that correct mathematically derived results stand on their own, and are therefore worth publishing. By now, I have realized that this is clearly not the case. Although physical intuition does not always (at first) go hand in hand with mathematically or experimentally derived results, one should always strive for a complete physical understanding behind the observed outcomes[2].

When it comes to scientific thinking, my supervisor has taught me to "*let go of the equality sign*", and think in terms of inequalities and counterexamples. Although I am still (and will always be) in a learning process, I noticed that this way of thinking is highly effective, as inequalities are typically much easier to find (but are very informative), and counterexamples immediately imply when a statement is incorrect. Furthermore, asking questions and having discussions provide an iterative procedure that naturally leads to deeper insight. Also here I definitely gained more experience, but clearly my ways of thinking are far from being perfect.

When it comes to subject specific knowledge I have learned a great deal about the Ising model, analytical techniques (e.g. the LEQ approximation), computer languages (e.g. Mathematica and Python), and computer simulations. Especially the former two brought me a lot of joy, and I am looking forward to learn more about the secrets of the simplistic, yet complex, Ising model.

[2] Unless the research is done in the field of pure mathematics.

References

1. Istrail S (2000) Statistical mechanics, three-dimensionality and np-completeness: I. universality of intracatability for the partition function of the ising model across non-planar surfaces (extended abstract). In: Proceedings of the thirty-second annual ACM symposium on theory of computing, ser. STOC '00. Association for Computing Machinery, New York, NY, USA, pp 87–96. [Online]. Available: https://doi.org/10.1145/335305.335316
2. Serra N, Di Carlo P, Rea T, Sergi CM (2021) Diffusion modeling of COVID-19 under lockdown. Phys Fluids 33(4):041903. [Online]. Available: https://doi.org/10.1063/5.0044061
3. Mello IF, Squillante L, Gomes GO et al (2021) Epidemics, the ising-model and percolation theory: a comprehensive review focused on covid-19. Physica A 573:125963. [Online]. Available: https://www.sciencedirect.com/science/article/pii/S0378437121002351
4. Machado D, Mulet R (2021) From random point processes to hierarchical cavity master equations for stochastic dynamics of disordered systems in random graphs: ising models and epidemics. Phys Rev E 104:054303. [Online]. Available: https://link.aps.org/doi/10.1103/PhysRevE.104.054303
5. Bethe HA (1935) Statistical theory of superlattices. Proc Math Phys Eng Sci 150(871):552–575. [Online]. Available: https://royalsocietypublishing.org/doi/abs/10.1098/rspa.1935.0122
6. Guggenheim EA (1935) The statistical mechanics of regular solutions. Proc Math Phys Eng Sci 148(864):304–312. [Online]. Available: https://royalsocietypublishing.org/doi/abs/10.1098/rspa.1935.0020
7. de With G (2013) Liquid-state physical chemistry: fundamentals, modeling, and applications. Wiley-VCH Verlag
8. Kikuchi R, Cahn JW (1962) Theory of domain walls in ordered structures–ii: pair approximation for nonzero temperatures J Phys Chem Solids 23(1–2):137–151. [Online]. Available: https://doi.org/10.1016/0022-3697(62)90067-7
9. Speck T, Reister E, Seifert U (2010) Specific adhesion of membranes: mapping to an effective bond lattice gas. Phys Rev E 82:021923. [Online]. Available: https://link.aps.org/doi/10.1103/PhysRevE.82.021923
10. Fenz SF, Bihr T, Schmidt D et al (2017) Membrane fluctuations mediate lateral interaction between cadherin bonds. Nat Phys 13(9):906–913. [Online]. Available: https://www.nature.com/articles/nphys4138
11. Meibohm J, Esposito M (2022) Landau theory for finite-time dynamical phase transitions. arXiv preprint arXiv:2205.10311. [Online]. Available: https://arxiv.org/abs/2205.10311
12. Külske C, Le Ny A (2007) Spin-flip dynamics of the curie-weiss model: loss of gibbsianness with possibly broken symmetry. Commun Math Phys 271(2):431–454. [Online]. Available: http://dx.doi.org/10.1007/s00220-007-0201-y
13. Ermolaev V, Külske C (2010) Low-temperature dynamics of the curie-weiss model: periodic orbits, multiple histories, and loss of gibbsianness. J Stat Phys 141(5):727–756. [Online]. Available: http://dx.doi.org/10.1007/s10955-010-0074-x
14. Meibohm J, Esposito M (2022) Finite-time dynamical phase transition in nonequilibrium relaxation. Phys Rev Lett 128:110603. [Online]. Available: https://link.aps.org/doi/10.1103/PhysRevLett.128.110603
15. Kikuchi R (1951) A theory of cooperative phenomena. Phys Rev 81:988–1003. [Online]. Available: https://link.aps.org/doi/10.1103/PhysRev.81.988
16. Mézard M, Sakellariou J (2011) Exact mean-field inference in asymmetric kinetic ising systems. J Stat Mech Theor Exp 2011(07):L07001. [Online]. Available: https://doi.org/10.1088/1742-5468/2011/07/l07001
17. Aguilera M, Moosavi SA, Shimazaki H (2021) A unifying framework for mean-field theories of asymmetric kinetic ising systems. Nat Commun 12(1):1197. [Online]. Available: https://doi.org/10.1038/s41467-021-20890-5
18. Reiter J (1991) Statics and dynamics of the two-spin-facilitated kinetic ising model. J Chem Phys 95(1):544–554. [Online]. Available: https://doi.org/10.1063/1.461455

19. Schulz M, Reineker P (1995) Interface growth in a two-spin facilitated kinetic ising model. Phys Rev B 52:4131–4137. [Online]. Available: https://link.aps.org/doi/10.1103/PhysRevB.52.4131
20. Godrèche C, Pleimling M (2018) Freezing in stripe states for kinetic ising models: a comparative study of three dynamics. J Stat Mech Theory Exp 2018(4):043209. [Online]. Available: https://doi.org/10.1088/1742-5468/aab67b
21. Gonzalez-Miranda JM, Garido PL, Marro J et al (1987) Nonequilibrium phase diagram of ising model with competing dynamics. Phys Rev Lett 59:1934–1937. [Online]. Available: https://link.aps.org/doi/10.1103/PhysRevLett.59.1934
22. Dickman R (1987) Kinetic phase transitions and tricritical point in an ising model with competing dynamics. Phys Lett A 122(9):463–466. [Online]. Available: https://www.sciencedirect.com/science/article/pii/037596018790867X
23. Szolnoki A (2000) Phase transitions in the kinetic ising model with competing dynamics. Phys Rev E 62:7466–7469. [Online]. Available: https://link.aps.org/doi/10.1103/PhysRevE.62.7466
24. Kumar M, Dasgupta C (2020) Nonequilibrium phase transition in an ising model without detailed balance. Phys. Rev E 102:052111. [Online]. Available: https://link.aps.org/doi/10.1103/PhysRevE.102.052111
25. Li YI, Cates ME (2020) Non-equilibrium phase separation with reactions: a canonical model and its behaviour. J Stat Mech Theory Exp 2020(5):053206. [Online]. Available: https://doi.org/10.1088/1742-5468/ab7e2d
26. Ouazan-Reboul V, Agudo-Canalejo J, Golestanian R (2021) Non-equilibrium phase separation in mixtures of catalytically active particles: size dispersity and screening effects. Eur Phys J E 44(9):1–10. [Online]. Available: https://doi.org/10.1140/epje/s10189-021-00118-6
27. Albano E, Binder K, Paul W (2000) Monte carlo studies of d= 2 ising strips with long-range boundary fields. J Phys. Condens Matter 12(12):2701. [Online]. Available: https://doi.org/10.1088/0953-8984/12/12/311
28. Lipowsky R, Kroll DM, Zia RKP (1983) Effective field theory for interface delocalization transitions. Phys Rev B 27:4499–4502. [Online]. Available: https://link.aps.org/doi/10.1103/PhysRevB.27.4499

Curriculum Vitae

Kristian Blom

Personal Information

Email address: kristian.blom@mpinat.mpg.de
Date of birth: 11-02-1996
Nationality: Dutch
Website: https://www.mpinat.mpg.de/godec

Education + Research Experience

2022–2023 **Postdoctoral Researcher in Math. bioPhys. group, Max Planck Institute for Multidisciplinary Sciences, Göttingen (DE)**
- Supervisor: *Dr. Aljaž Godec*

2018–2022 **Ph.D. in Math. bioPhys. group, Max Planck Institute for Multidisciplinary Sciences, Göttingen (DE)**
- Supervisor: *Dr. Aljaž Godec*
- Title: *Pair-correlations in many-body systems*
- Grade: *Summa cum laude (with highest honors)*

2017–2018 **M.Sc. in Mathematical and Theoretical Physics, University of Oxford, Oxford (UK)**
- Coursework: *general relativity, quantum field theory, string theory, and cosmology*

2015–2017 **Honours Program Nanobiology, TU Delft and Erasmus University, Delft (NL)**

K. Blom, *Pair-Correlation Effects in Many-Body Systems*, Springer Theses,
https://doi.org/10.1007/978-3-031-29612-3

	• For students in the top 5% of the year • Coursework: *complex function theory, special relativity, presenting & debating*
2014–2017	**B.Sc. in Nanobiology, TU Delft and Erasmus University, Delft (NL)** • Thesis title: *From modeling target recognition towards a target prediction algorithm for RNA-guided nucleases* • Coursework: *analysis 1–3, differential equations, classical mechanics, quantum mechanics, statistical physics, signal theory, optics, chemistry, and bioinformatics* • Minor in theoretical astronomy • Grade: *Cum laude*

Publications

2023–2024	• Blom, K., Godec, A., " *Global Speed Limit for Finite-Time Dynamical Phase Transition in Nonequilibrium Relaxation*" *ArXiv:2209.14287* • Blom, K., Ziethen, N., Zwicker, D., Godec, A., " *Thermodynamically Consistent Phase-Field Theory Including Nearest-Neighbor Pair Correlations*" *Phys. Rev. Res.* ***5.****1 (2023)* • Blom, K., " *Een kritisch punt in de bindingssterkte van cellen*" *NTvN* ***88.*** *4 (2022)*
2021–2022	• Blom, K., and Godec, A., " *Criticality in Cell Adhesion*" *Phys. Rev. X* ***11.*** *3 (2021)*
2017–2018	• *Future Engineering in Asia Honours Research.*

Conference Contributions

2022–2023	• **(Post)Modern Thermodynamics, Luxembourg (LUX), Poster** Poster title: *Global Speed Limit for Finite-Time Dynamical Phase Transition in Nonequilibrium Relaxation* • **Entropy and the Second Law of Thermodynamics, Bad Honnef (DE), Poster** Poster title: *Tuning Kinetics of Magnetization Reversal in an Irreversible Ising Model*
2021–2022	• **The** 4th **World Laureates Forum 2021, Shanghai (CN), Poster + Video** Talk title: *Criticality in the Ising model and cell adhesion* • German Physics Society (DPG) Spring Meeting 2021, (DE), Talk Talk title: *Criticality in the Mechanical Regulation of Cell Adhesion*
2020–2021	• Biophysical Society (BPS) Meeting 2020, San Diego (US), Talk Talk title: *Collective Mechanochemical Effects in Microtubule Dynamics: Theory and Simulations*

2019–2020 • Biology and Physics Confront Cell-Cell Adhesion, Aussois (FR), Poster
Poster title: *Cellular Adhesion Dynamics Under an External Force*

Teaching Experience

04/2020–07/2020	Teaching assistant *Quantum Mechanics 1*, Physics (B.Sc.), University of Göttingen
10/2019–03/2020	Teaching assistant *Advanced Methods in Statistical Mechanics*, International Max Planck Research School (IMPRS)
10/2018–01/2019	Teaching assistant *Cosmology*, Physics (M.Sc.), University of Göttingen
03/2017–05/2017	Teaching assistant *Biophysics*, Nanobiology (B.Sc.), TU Delft
11/2016–03/2017	Teacher *System & Signals*, Athena Studies, Delft
11/2016–01/2017	Teaching assistant + examinator *Signals & Systems*, Applied Physics (B.Sc.), TU Delft
04/2016–06/2016	Teaching assistant *Analysis 3 + Linear Algebra 1*, Nanobiology (B.Sc.), TU Delft

Skills

Programming/Software

Language	Level	Experience
MATLAB/Octave	•••	Taught during my bachelor with a 9.5/10 as final mark
Python	•••	Used for bachelor thesis and during my Ph.D. studies
JavaScript	••	Taught during my bachelor to create a diffusing N-body simulator
Mathematica	•••	Taught during my master and used during my Ph.D. studies
Blender	•	Used to produce various animations during my Ph.D. studies

Languages

Language	Speaking	Writing	Reading	Comments
Dutch	•••	•••	•••	Native language
English	•••	•••	•••	Cambridge Advanced English (199) /ETF TOEFL (101)
German	••	••	••	Certificate obtained in level B2.2

Extracurricular Activities

2022–2023	• Student representative of the International Max Planck Research School (IMPRS)
2021–2022	• Organization of the International Max Planck Research School (IMPRS) retreat 2021
2019–2020	• Organization of scientific conference Third Infinity 2020 • Speaker at 4th Göttingen Science Night
2017–2018	• Member of the CCG: Oxford Consultative Committee for Graduates • Chairman Tokyo tour commission for the Honours Program • Speaker for 'Museum Youth University' about the origin of aging in cells • Participant summer school in 'Theoretical Physics' at Utrecht University
2015–2016	• Chairman faculty student council Applied Sciences at the TU Delft • Chairman CERN tour commission for the Honours Program at the TU Delft • Chairman Erasmus MC Summer School in Nanobiology • Speaker for the lecture series 'Mysteries of the unseen world' Science Centre Delft • Participant summer school 'Physics of the Climate System' at Utrecht University

Scholarships

2018–2021	• For my Ph.D. studies at the Max Planck Insititute: IMPRS excellence fellowship
2016–2017	• For my M.Sc. at the University of Oxford: Genootschap Noorthey and Vreedefonds (each 6000€), Stichting Vrijvrouwe van Renswoude and Hendrik Muller Fonds (each 4000€). Total: 20,000€

Prizes/Awards

2023–2024	• Institute of Physics (IOP) Outstanding Reviewer Award for Journal of Physics A: Mathematical and Theoretical
2022–2023	• Institute of Physics (IOP) Trusted Reviewer • Dissertation prize Springer Thesis Award • Winner of the Dutch Journal for Physics essay competition, Twente (NL)
2021–2022	• Accepted to the 4th World Laureates Forum in Shanghai (CN) • Winner of the 2nd Poster Prize at the summer school Fundamental Problems in Statistical Physics XV, Bruneck (IT)

2019–2020 • Winner of the Biophysical Journal Poster Competition at the BPS Thematic Meeting 'Biology and Physics Confront Cell-Cell Adhesion', Aussois (FR)

MIX
Papier aus verantwortungsvollen Quellen
Paper from responsible sources
FSC® C105338

If you have any concerns about our products,
you can contact us on
ProductSafety@springernature.com

In case Publisher is established outside the EU,
the EU authorized representative is:
Springer Nature Customer Service Center GmbH
Europaplatz 3, 69115 Heidelberg, Germany

Printed by Libri Plureos GmbH
in Hamburg, Germany